RICH EARTH

Alaska's Mineral Industry

ALASKA GEOGRAPHIC / Volume 22, Number 3

To teach many more to better know and more wisely use our natural resources...

POSTMASTER: Send address changes to
ALASKA GEOGRAPHIC®
P.O. Box 93370
Anchorage, Alaska 99509-3370

PRINTED IN U.S.A.

ISBN: 1-56661-027-3

Price to non-members this issue: $19.95

COVER: *For many decades the floating gold dredge, such as this one abandoned near Nome, symbolized Alaska's mineral industry. (Jon R. Nickles)*

PREVIOUS PAGE: *Ruins of the historic Alaska-Juneau Mine dominate this hillside above Gastineau Channel near Juneau. (Jon R. Nickles)*

FACING PAGE: *The gold pan held the hopes of individual prospectors and miners, and today lures tourists to try their luck at various recreational gold-panning sites throughout the state. (Harry M. Walker)*

ALASKA GEOGRAPHIC® (ISSN 0361-1353) is published quarterly by The Alaska Geographic Society, 639 West International Airport Road, Unit 38, Anchorage, AK 99518. Second-class postage paid at Anchorage, Alaska, and additional mailing offices.

THE ALASKA GEOGRAPHIC SOCIETY is a non-profit, educational organization dedicated to improving geographic understanding of Alaska and the North, putting geography back in the classroom and exploring new methods of teaching and learning.

MEMBERS RECEIVE *ALASKA GEOGRAPHIC®*, a high-quality quarterly publication that devotes each issue to monographic, in-depth coverage of a northern region or resource-oriented subject.

MEMBERSHIP in The Alaska Geographic Society costs $39 per year, $49 to non-U.S. addresses. ($31.20 of the membership fee is for a one-year subscription to *ALASKA GEOGRAPHIC®*.) Back issues are also available. To order or request a free catalog of available back issues, write to: The Alaska Geographic Society, Box 93370, Anchorage, AK 99509-3370; phone (907) 562-0164, fax (907) 562-0479, e-mail: akgeo@anc.ak.net.

SUBMITTING PHOTOGRAPHS: Please write for a list of upcoming topics or other specific photo needs and a copy of our editorial guidelines. We cannot be responsible for unsolicited submissions. Submissions not accompanied by sufficient postage for return by certified mail will be returned by regular mail.

CHANGE OF ADDRESS: The post office does not automatically forward *ALASKA GEOGRAPHIC®* when you move. To ensure continuous service, please notify us at least six weeks before moving. Send your new address and your membership number or a mailing label from a recent *ALASKA GEOGRAPHIC®* to: Alaska Geographic Society, Box 93370, Anchorage, AK 99509-3370.

COLOR SEPARATIONS: Graphic Chromatics

PRINTING: The Hart Press

EDITOR'S NOTE: In mid-1995, Alaska was enjoying a boom in its non-fossil-fuel mining industry. It has been decades since Alaskans have been able to count so many major mines in the works and such viable prospects for more mines. This issue focuses on the resurgence in the hardrock and placer mining industry as it applies to the state's non-fossil-fuel resources. Readers will find no oil, gas or coal, but they will find gold and other significant minerals, including strategic and industrial minerals. Combined, these minerals generated a $507.5 million harvest in 1994 from Alaska's rich earth.

ABOUT THIS ISSUE: This issue couldn't have been done without the help of Thomas K. Bundtzen with the Alaska Geological Survey in Fairbanks, who shares graciously his abundant knowledge of historical and contemporary mining in Alaska. We appreciate his help in gathering information, his contribution of photos and his review of the manuscript. We also acknowledge the information provided by Steve Borell, executive director of the Alaska Miners Association, and Ralph Eastman, of Cominco Ltd. in Vancouver, British Columbia. We appreciate the information, insights and mining lingo lesson provided by Richard Flanders, a consulting geologist from Fairbanks. We also thank members of the U.S. Geological Survey for providing facts, photographs, interviews, comments and other assistance, especially Bob Chapman and Helen Reiser, in Menlo Park, Calif., and Elizabeth Bailey, Julie Dumoulin, Don Grybeck, Tom Hamilton, Marti Miller, Tom Miller, Steve Nelson and Jill Schneider, all in Anchorage.

The Library of Congress has cataloged this serial publication as follows:

Alaska Geographic. v.1-
[Anchorage, Alaska Geographic Society] 1972-
v. ill. (part col.). 23 x 31 cm.
Quarterly
Official publication of The Alaska Geographic Society.
Key title: Alaska geographic, ISSN 0361-1353.

1. Alaska—Description and travel—1959-
—Periodicals. I. Alaska Geographic Society.

F901.A266 917.98'04'505 72-92087

Library of Congress 75[79112] MARC-S.

Contents

DON J. MILLER
DON J MILLER II
USGS
DON J MILLER II
USGS

Marking the Map's Void Spaces

USGS Geologists in Alaska

By Richard P. Emanuel

Editor's note: *A frequent contributor to* ALASKA GEOGRAPHIC®, *Dick Emanuel knows Alaska's geological environment well. He has worked for the U.S. Geological Survey and focused much of his effort on the geology and hydrology of Mount Spurr.*

Have you marked the map's void spaces...?
—Robert Service, from the poem,
"The Call of the Wild"

As the 20th century draws to a close, the future of the U. S. Geological Survey in Alaska is uncertain. But whatever its future, the agency's past is marked by more than a century of notable achievements by a host of remarkable individuals. Survey scientists have helped to open Alaska and have had a hand in shaping the state.

Since the first USGS scientist was dispatched to Alaska in 1889, the field and practice of geology have greatly changed. Through war, Depression and peace, the Survey has experienced famine and feast. The feasting is over for now. The USGS, like other federal agencies, is today enmeshed in a process of self-examination and "reinvention." One possible new direction for Survey geologists in Alaska is being pursued by a team assembled by Marti Miller, of the Branch of Alaskan Geology.

The Don J. Miller II *was the last of three ships operated by the U.S. Geological Survey for mapping in coastal areas of Alaska, principally in Southeast and Prince William Sound, but also off the Alaska Peninsula. The ships provided a dry, mobile base for geologists and a heliport. The* Stephen R. Capps, *named for a Survey geologist who worked in Alaska from 1908 through 1936, operated from 1957 through 1963. The* Don J. Miller, *operated from 1964 through 1968, was named for a Survey geologist who worked in Alaska from 1942 until his drowning death in 1961. The* Don J. Miller II, *operated from 1969 through 1985, was a 118-foot converted tug. (Marti Miller)*

Miller has spent a decade mapping geology and studying mineral resources in southwestern Alaska. Recently, what began for her as a traditional assessment of mineral resources took on an environmental dimension.

That Miller is helping the Branch of Alaskan Geology explore in relatively untracked terrain seems fitting. She is a second-generation Survey geologist. Her father, Don J. Miller, worked for the Branch of Alaskan Geology from 1942 to 1961. Don Miller is best known for his work in Lituya Bay, where he studied evidence that gigantic waves wrack the bay from time to time. One such wave struck in 1936, and in the mid-1950s, Miller studied its effects. On July 9, 1958, an even greater wave was loosed when an earthquake sent a tremendous rockslide thundering into Lituya Bay. Two fishing boats in the outer bay were sunk, and two people were killed. Within days, Don Miller flew in to document the devastation. The formerly dense rain forest had been swept clean from the rocky shoreline to elevations greater than 1,700 feet. The results of Miller's research were presented in *Giant Waves in Lituya Bay, Alaska,* (USGS Professional Paper 354-C), in 1960.

The year after Miller's Lituya Bay work appeared, he and an assistant drowned while conducting field work in Alaska. Miller's backpack was found on a streambank, suggesting he may have removed it before attempting to rescue his comrade. Miller was 42. He was memorialized by his colleagues with the naming of the Don Miller Hills, a range of low mountains between Bering Lake and Controller Bay, 50 miles southeast of Cordova.

Marti Miller followed her father's footsteps after a while. In 1975, when she was

In 1984, Marti Miller of the USGS, Tom Bundtzen and Greg Laird of the Alaska Division of Geological and Geophysical Surveys, and Bob Betts, an archaeologist with the University of Alaska Museum in Fairbanks, spent 30 days working in southwestern Alaska with a pack train of seven horses. Outfitter Mark Meekins, who owned the horses, and Dee Larson, a cook, handled the animals and set up camp for the scientists each day. The party made a circuit of about 200 miles out of Flat, at a cost about equal to the cost of 10 days of helicopter support, according to Miller. "We covered about the same amount of ground as we would have by helicopter (in 10 days), but we got far more detail," Miller says. "It was the dream season for me." Here, wrangler Meekins leads some of the horses across Bonanza Creek. (Marti Miller, U.S. Geological Survey)

20, she got a summer job as a cook in a USGS field camp in Alaska. "That was really what made me decide to go into geology," she says. "I was attracted to the lifestyle."

Miller obtained a master's degree from Stanford University and landed a job with the Branch of Alaskan Geology in late 1980. She moved to Alaska in early 1981, and worked in Prince William Sound before turning her attention to southwestern Alaska.

In 1993, Miller began work in the area covered by the 1:250,000-scale topographic map called the Sleetmute quadrangle. Her previous work in the Iditarod quadrangle, to the north, had focused largely on gold. But a belt of mercury deposits runs through southwestern Alaska too, and the Sleetmute quad is best known for mercury. In 1838, Russian traders had found lodes of cinnabar, a reddish mercury-sulfur mineral, in the region. The first mercury mines in southwestern Alaska opened in 1909.

The region's biggest mercury producer,

the Red Devil Mine, was discovered in 1933. A 10-year-old berry picker found pieces of high-grade cinnabar on a hillside overlooking the Kuskokwim River, six miles northwest of Sleetmute. Red Devil went into production during World War II, to supply mercury needed for electrical instruments, munitions, pharmaceuticals and other things. Red Devil has produced some 36,000 flasks of the stuff, each flask holding 76 pounds of the silver-white liquid. The mine's 2.7 million pounds of the metal represent nearly 90 percent of all mercury mined in Alaska.

In the Sleetmute quad, as in her Iditarod work, Miller teamed with John Gray, a Survey geochemist based in Denver. In 1988, reconnoitering mercury and gold deposits in the Sleetmute area, the pair had visited an abandoned mine on Cinnabar Creek, just outside the Sleetmute quad boundary. The scientists were drawn by reports that ore at the site contained not just cinnabar but native mercury, the element in its pure form, a heavy liquid at normal temperatures.

"It's unusual to have native mercury," Miller says, "and these particular ore samples are stockpiled at the old mine. All you have to do is bang a sample on your thigh and little beads of mercury come popping out. Not surprisingly, the creek draining that area has local pools of mercury on the bottom."

The mine visit impressed Miller and Gray and they pondered the environmental implications of what they had seen. Mercury is poisonous, and they wondered if background levels of the toxic metal might be dangerously high in places in the mercury belt. "Perhaps we ought to be thinking about this as a possible public health problem," Miller recalls musing. "We're going to be out there anyway. It would behoove us to collect whatever samples we might need for evaluating mercury as a potential problem."

Environmental science is nothing new for the USGS, but it has not been a central interest of the Branch of Alaskan Geology. More often, environmental concerns have been the province of geochemists and hydrologists in other parts of the Survey. Still, opportunity counts for much in science, and Miller was on the scene. And in the 1990s, in Alaska as elsewhere, environmental questions are increasingly often asked of geologists.

Miller and Gray shared concern about environmental mercury, but as project geochemist, Gray took the lead in checking for contamination. Mercury exists in the environment in various forms but the most toxic is called methylmercury. Micro-organisms and natural chemical processes create methylmercury from other, less toxic forms. When small fish and other aquatic organisms take it up, most methylmercury remains stored in their bodies. Larger fish may eat smaller fish and concentrate the mercury in their livers and other organs. At the top of the food chain, bears or people may ingest contaminated fish and concentrate the methylmercury in turn. Kidney or brain damage may result. Fetuses and young children are especially vulnerable.

Sampling for environmental mercury can be tricky. Some forms of mercury readily

BELOW: *Mercury ore at Cinnabar Creek contains not only the reddish, mercury-sulfide mineral cinnabar, but also native mercury. Liquid beads of pure mercury pop out of ore samples when they are struck. (Marti Miller, U.S. Geological Survey)*

RIGHT: *Red Devil Mine, discovered by accident by a 10-year-old berry picker in 1933, has produced nearly 90 percent of all mercury mined in Alaska. The mine, near Sleetmute, is currently inactive. (Marti Miller, U.S. Geological Survey)*

CHUKCHI SEA
Barrow
ARCTIC OCEAN
FLAXMAN ISLAND
Prudhoe Bay
CAPE LISBURNE
Point Hope
DELONG MOUNTAINS
LIK
RED DOG
Colville River
Dalton Highway
Noatak River
BROOKS RANGE
Kivalina
BAIRD MOUNTAINS
Kotzebue Sound
Kotzebue
Kobuk River
SILVERADO
Wiseman
Shishmaref
Ambler
Bornite
Kobuk
Bering Strait
Porcupine River
Fort Yukon
Koyukuk River
Northwest Territories
Mackenzie River
Teller
SEWARD PENINSULA
Yukon River
Circle
Livengood
Nome
Council
BIG HURRAH
Solomon
Golovin
Steese Highway
Tanana
FORT KNOX
RYAN LODGE
TRUE NORTH
Fairbanks
Eagle
Manley
Ruby
Nulato
Tolovana
Yukon Territory
Norton Sound
KAIYUH MOUNTAINS
Fortymile
Chicken
George Parks Highway
Dawson City
CANADA
Tanana River
St. Michael
Healy
Telida
Kantishna
Tok
BERING SEA
Ophir
Takotna
Cantwell
VALDEZ CREEK
Fort Selkirk
Mount McKinley
Medfra
Nikolai
McGrath
Tok Cutoff
Iditarod
Flat
Vinasale
NIXON FORK
VON FRANK MT.
ALASKA RANGE
Denali Highway
NABESNA
Alaska Highway
Klondike Loop
KUSKOKWIM MOUNTAINS
Richardson Highway
Talkeetna
INDEPENDENCE
WRANGELL MOUNTAINS
RED DEVIL
POINT MACKENZIE
Glenn Highway
Whitehorse
Alaska Highway
Kuskokwim River
Palmer
Chitina
Mount Spurr
Anchorage
Valdez
Bethel
Nyac
Haines Highway
CHILKOOT PASS
Klondike Highway
Kenai
Cordova
Mount St. Elias
British Columbia
WHITE PASS
Cook Inlet
KENAI PENINSULA
Seward
Katalla
Yakutat
Klukwan
Skagway
Haines
PEBBLE COPPER
Iliamna
Homer
Bering Glacier
Malaspina Glacier
DON MILLER
Goodnews Bay
Dillingham
RED TOP
Seldovia
Prince William Sound
Lituya Bay
Brady Glacier
KENSINGTON & JUALIN
Juneau
ALASKA-JUNEAU
TREADWELL
GREENS CREEK
GLACIER BAY
PRIBILOF ISLANDS
BRISTOL BAY
Naknek
GULF OF ALASKA
CHICHAGOF ISLAND
ADMIRALTY ISLAND
Sitka
Kodiak
Petersburg
Wrangell
Hyder
ALASKA PENINSULA
QUARTZ HILL
Tokeen
SALT CHUCK
Ketchikan
PRINCE OF WALES ISLAND
BOKAN MT.
KASAAN PENINSULA
Hetta Inlet
ALASKA
MINES/MINING PROSPECTS
MILES
50 100 150 200 250
(©1995, The Alaska Geographic Society; map by Kathy Doogan)
BOGOSLOF ISLAND
APOLLO
SHUMAGIN ISLANDS
Unalaska

evaporate, so soil or water samples must be carefully sealed. Since fish may concentrate mercury above its "background" level in streams or lakes — and since people eat fish — to check for contamination, it is often best to directly test fish.

During summer 1993, in addition to mapping the geology and collecting geochemical samples needed to assess the area's mineral resource potential, Miller and Gray collected samples around known mercury deposits to test for contamination. Gray collected water and sediment from streams. He used a gold pan to concentrate heavy minerals in stream sediment, and sampled the concentrate. At times, he unlimbered a fly rod and sampled arctic grayling. These he later tested for mercury.

In late 1993, while Gray was still at work in his Denver lab, Miller got a call from a Bureau of Land Management geologist. "He was concerned about mercury," Miller says. "BLM has responsibility for the Red Devil Mine site."

The BLM's concern about mercury had been heightened the year before, when Greenpeace activists sampled soil and water at the abandoned Red Top Mine, near Dillingham. Native mercury was visible in the soil at the site, beneath an old retort. Cinnabar ore had been heated in the retort to drive off mercury vapor, which condensed as it cooled and was collected in flasks. Local Natives, concerned about the abandoned mine, had contacted Greenpeace. The environmental group investigated the site and released its findings to the state Department of Environmental Conservation — and to news media. The Red Top site had been scheduled for cleanup, but the BLM was anxious to avoid similar imbroglios elsewhere.

Marti Miller is a second generation USGS geologist. She has worked in southwestern Alaska for the last decade. (Courtesy of Marti Miller, U.S. Geological Survey)

In January 1994, the BLM convened an interagency meeting in Anchorage to talk about mercury. Eight state and federal agencies sent representatives. Miller described the geology of southwestern Alaska's mercury belt, and Gray presented the results of his geochemical sampling. Elizabeth Bailey, a colleague of Miller's in the Branch of Alaskan Geology, also attended. Bailey was interested in mercury accumulation in plants, and wanted to research the topic for a master's degree in environmental science. Mercury is fairly well studied in the aquatic food chain, but little was known about whether or how it accumulates in plants.

As a result of the meeting, an agreement was struck between the USGS and the U.S. Bureau of Mines to cooperate in mercury studies in the Sleetmute area during summer 1994.

Mike Balen, a Bureau of Mines mining engineer, inventoried abandoned mine dumps at sites along the Kuskokwim River. He described the ore dumps, observed the effluent draining from the sites and sampled the soil and water. Miller and Bill Keith, of the USGS, and Tom Bundtzen, of the Alaska Division of Geological and Geophysical Surveys, mapped the geology of the Sleetmute quad with special attention to the mine sites. Gray continued his geochemical sampling at mine sites and throughout the quad, and Bailey collected plant and soil samples for her research.

Gray has found that fish downstream from abandoned mines contain more mercury than fish elsewhere, but not enough to pose a health threat. Fish with the highest mercury levels, grayling from Cinnabar Creek, average 0.43 parts per million mercury, less than half the 1.00 ppm limit set by the Food and Drug Administration. The maximum value found was 0.60 ppm, of which five-sixths was methylmercury. None of the anadromous fish species (salmon) caught in the Kuskokwim River drainage had any mercury contamination.

Bailey's effort is also starting to bear fruit. To begin with, she says, "I wanted to know whether plants take up methylmercury.

A curious horse named Steve investigates archaeologist Bob Betts' morning shaving ritual, as wrangler Mark Meekins looks on. (Marti Miller, U.S. Geological Survey)

There is almost no scientific literature on methylmercury in vegetation." Bailey sampled alders and other plants at the Red Devil and Cinnabar Creek mines, and in control areas away from known deposits. Plants from the mine sites contain more mercury. "The highest concentrations are in leaves," Bailey adds. Alder leaves at the Cinnabar Creek Mine contain an average of 0.34 ppm mercury, nearly three times the level in alder leaves away from the mine site. The highest concentration found, in leaves from an alder growing in soil with native mercury, was 0.97 ppm. Mercury levels were lower in seed pods and fruit and lower still in stems. Some of the mercury was in the form of toxic methylmercury, but it represents only a small fraction of the total mercury found in plant and soil samples. Background levels, some distance from the mines, are generally less than 0.05 ppm.

The danger posed by mercury in plants remains problematic. In theory, a moose might browse on contaminated alders, and a bear or human might eat the moose. But no one knows how efficiently a moose concentrates mercury, and in any case, a moose might range across a wide area, diluting the contaminated food. Its total mercury load might not be significant. "I've got a lot of questions," Bailey says. But every answer she uncovers adds new information.

"The results so far indicate that it's probably not a huge problem," Miller says in summary. "But it's something that needs to be considered. And Sleetmute is just part of southwest Alaska. We'd like to see this type of study encompass all of the mercury belt."

Bailey agrees and broadens the view. "I think we should do more to predict the behavior of other metals at mine sites. How does the weathering of spoil piles affect the movement of these metals? How do they behave in plants and animals, especially in known food sources, like berries?"

Despite the start made by Miller, Gray and Bailey, funding for continued work is scarce. The Branch of Alaskan Geology and the USGS as a whole are facing serious budget cutbacks and possible layoffs. Indeed, the Branch of Alaskan Geology, now 92 years old, may not see its centennial.

The Days of Dall

The U.S. Geological Survey was established by Congress in 1879, 12 years after the United States acquired Alaska. Even then, the man destined to become the Survey's first Alaska expert had already been to Alaska. The land was still in Russian hands when William Healey Dall, the naturalist for whom Dall sheep were named, first broached the boreal wilderness.

Dall was born in 1845 in Boston, to an ardent feminist and a Unitarian minister. He studied under Louis Agassiz, Harvard's inspiring professor of zoology and geology. At age 18, Dall headed west to Chicago, where he took a job as a clerk with the Illinois Central Railroad.

In Chicago, Dall met Robert Kennicott, a naturalist at the Chicago Academy of Sciences who had visited Alaska. When the Western Union Telegraph Co. decided to build a telegraph line across Bering Strait to Asia and Europe, Kennicott was hired to take charge of the Russian-American segment. He convinced the company to sponsor a small Scientific Corps as part of the necessary exploration. Among the half-dozen scientists who joined him in Alaska in 1865 was William Dall.

The Western Union expedition to Alaska was ill-fated. In May 1866, Kennicott, 30, suffered an apparent heart attack and died near Nulato. Dall succeeded him as chief of the Scientific Corps and explored up the Yukon River as far as Fort Yukon. But when a trans-Atlantic cable to Europe was completed in 1867, Western Union summarily suspended the Alaska expedition.

Still, Dall stayed on in Alaska. In February 1868, he learned that the United States had bought Alaska from Russia a few months earlier. He took great pleasure in raising the Stars and Stripes in front of the fort at Nulato. In August 1868, Dall sailed from St. Michael. At the Smithsonian Institution in Washington, D.C., he studied and catalogued some 4,550 rock, plant and animal specimens collected in Alaska. He presented his experiences and findings in a book, *Alaska and Its Resources* (1870).

Dall's book established him as the leading scientific authority on Alaska. Between 1871 and 1880, he made reconnaissance surveys with the U.S. Coast Survey, later renamed the Coast and Geodetic Survey. He ranged along coastal areas from Sitka westward to Attu, and northward to the Pribilof Islands and along the Arctic coast nearly to Point Barrow.

In 1884, Dall joined the U.S. Geological Survey, then five years old. He was attached to the U.S. National Museum as a paleontologist, an association he maintained until his death in 1927.

In establishing the USGS, Congress had charged it with "examination of the geological structure, mineral resources, and

William Healey Dall (1845-1927) was among the first American scientists in Alaska. Dall, shown here at age 33, went to Alaska in 1865, while it was still in Russian hands, as part of the Western Union Telegraph Expedition. He traveled in the Yukon drainage for three years, collected rock, plant and animal specimens and wrote a book which established him as the leading scientific authority on Alaska. Dall made reconnaissance surveys of Alaska's coast for the U.S. Coast Survey, later the Coast and Geodetic Survey, from Sitka to Attu, northward to the Pribilof Islands and later nearly to Point Barrow. He joined the USGS in 1884. In 1895, Dall, G.F. Becker and C.W. Purington reconnoitered gold and coal deposits along the coast from Sitka to Unalaska. (U.S. Geological Survey)

Geologist Israel Cook Russell (1852-1906) was the first USGS scientist sent to Alaska. In 1889, Russell served as "geological attaché" to a U.S. Coast and Geodetic Survey party reconnoitering the U.S.-Canada boundary by way of the Yukon and Porcupine rivers. In 1890 and 1891, Russell explored and mapped Malaspina Glacier, Yakutat Bay and the Mount. St. Elias region. (U.S. Geological Survey)

products of the national domain." Preoccupied with the Lower 48, the agency did not send a scientist to Alaska until 1889. Then, Israel Cook Russell was assigned as a "geological attaché" to a U.S. Coast and Geodetic Survey party reconnoitering the Alaska-Canada boundary.

Russell and the survey party traveled up the Yukon River to Fort Yukon, then ascended the Porcupine River and returned. They journeyed upriver from Fort Yukon and finally crossed over Chilkoot Pass to the sea. Along the way, Russell observed the geology and confirmed Dall's conclusion that the Yukon Valley had not been covered with glaciers during the Pleistocene Epoch, the last great ice age.

In 1890 and 1891, with joint funding from the USGS and National Geographic Society, Russell led explorations of Mount St. Elias, then thought to be North America's highest peak. Russell was particularly interested in glaciers and explored giant Malaspina Glacier, as well as many smaller alpine glaciers. The party mapped the topography of more than 1,000 square miles during its two field seasons.

1891 saw another USGS geologist in Alaska, C. Willard Hayes. Hayes joined an expedition organized by former Army explorer Frederick Schwatka, a private exploit funded by a syndicate of some 50 newspapers. Schwatka had asked Survey director John Wesley Powell to supply a geologist, and Hayes had volunteered.

The Schwatka expedition ascended the Taku River, northeast of Juneau, and crossed the divide to Teslin Lake, in Canada. Unfolding portable canoes, they paddled down the Teslin River to the Yukon. From Fort Selkirk, on the Yukon, the group ascended the White River to its headwaters, portaged to the upper Chitina River and descended the Chitina and Copper rivers to the coast. Hayes made observations along the way on topography, geology, mineral resources including copper deposits, volcanoes, glaciers and vegetation. His hastily drawn map was later found "remarkably complete and accurate."

LEGEND

1889 *I.C. Russell with U.S. Coast & Geodetic Survey party, up Yukon and Porcupine rivers, then upper Yukon drainage to Chilkoot Pass; 2,000 miles*

1891 *C.W. Hayes with Frederick Schwatka expedition, up Taku River to Teslin Lake and upper Yukon to Fort Selkirk, up White River to Skolai Pass, down Chitina and Copper rivers; 800 miles*

1895 *W.H. Dall, G.F. Becker and C.W. Purington, by ship from Sitka to Unalaska including Kodiak and lower Cook Inlet and volcanic islands Augustine and Bogoslof; more than 2,000 miles*

1896 *J.E. Spurr, F.C. Schrader and H.B. Goodrich, over Chilkoot Pass to Yukon River and downriver to St. Michael, with stops in the Fortymile, Birch Creek, Eagle and Rampart gold districts; 2,000 miles*

1898 *J.E. Spurr and W.S. Post, from Tyonek up Yentna and Skwentna rivers, across Alaska Range to the Kuskokwim drainage, downriver to Kuskokwim Bay, up Kanektok River toTogiak Lake, Nushagak, and across Alaska Peninsula via Naknek Lake portage to Katmai; 1,400 miles*

1898 *W.C. Mendenhall with U.S. Army expedition of Capt. E.F. Glenn, from Seward via Resurrection Trail to Turnagain Arm, via Crow Pass to Knik, up Matanuska River and across Alaska Range to Tanana River; 800 miles*

1901 *W.C. Mendenhall and D.L. Reaburn, from Fort Hamlin on Yukon River by way of Dall, Kanuti, Alatna and Kobuk rivers to Kotzebue Sound; 1,200 miles*

1901 *W.J. Peters and F.C. Schrader, from Yukon River across Brooks Range at Anaktuvuk Pass to Arctic Ocean and along coastline to Barrow, by ship to Cape Lisburne; 800 miles to Barrow*

1902 *A.H. Brooks, D.L. Reaburn and L.M. Prindle, from Tyonek across the Alaska Range through Rainy Pass, northeast 200 miles along base of Alaska Range past McKinley to Nenana drainage, northward across Tanana River to Rampart; 800 miles*

1924 *P.S. Smith, J.B. Mertie, G. FitzGerald and R.K. Lynt, from Nenana to Tanana, north across Brooks Range through Survey Pass to Killik River in winter; after breakup by canoe in two parties to arctic coast and Barrow; by ship to Nome; 700 miles to Barrow*

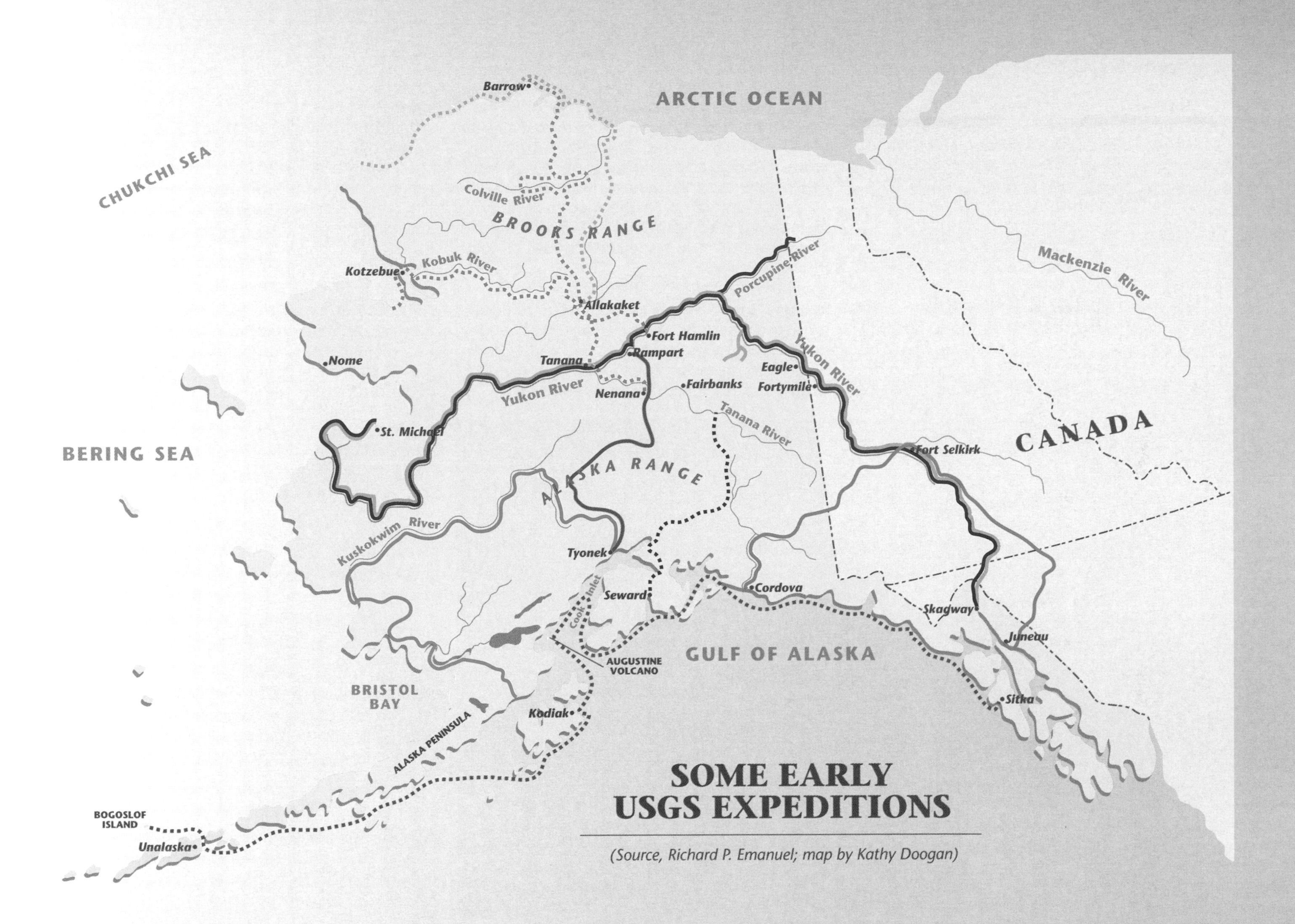

SOME EARLY USGS EXPEDITIONS

(Source, Richard P. Emanuel; map by Kathy Doogan)

Ascendancy of the Geological Survey

By the close of the century the major features of Alaska's portrait had been sketched, if not drawn in detail, and there were few problems of gross geography left unsolved; this success mainly resulted not from work done by the Army, but from the labors of a small corps of USGS explorers.
—Morgan B. Sherwood, Exploration of Alaska *(1965)*

The year 1895 saw the USGS' first independent undertakings in Alaska, a study of gold and coal resources. William Dall and two other scientists spent a month in Southeast Alaska aboard a naval steamer, then took the mail boat to Kodiak where they chartered a small tug. They reconnoitered the coast of Kodiak Island, lower Cook Inlet, the Alaska Peninsula and the Aleutians as far west as Unalaska. They did not confine their studies to coal and gold. In lower Cook Inlet, one of the scientists climbed the island volcano, Augustine. In the Aleutians, they visited Bogoslof Island, which had reportedly risen from the sea in volcanic eruptions just 99 years earlier.

In 1896, Josiah E. Spurr led two other geologists on a survey of upper Yukon gold fields. The party crossed Chilkoot Pass and descended the Yukon to the Fortymile River, near Eagle. Spurr and his party visited active mines and prospects and made their way down the Yukon to study gold in the Birch Creek area and around Rampart. The geologists were in Alaska when gold was struck in the Fortymile District and in Canada's Klondike, although the size of the strikes was not immediately clear. Spurr's report when he returned to Washington, D.C., however, was enough to prompt the Survey's director to boost his budget request for the next year's Alaska work from $2,500 to $25,000. Congress did not grant that sum, however, until 1900.

During most of the 19th century, national interest in Alaska had been less than keen. But as the century drew to a close, gold fueled a feverish curiosity about the dim and distant north. In 1895, Alaska produced more than $2.5 million in gold, much of it from Yukon Valley placer mines. Seven steamers operated on the Yukon in 1896, serving a mining population about 1,000 strong.

To guide them through the wild and unfamiliar countryside, miners needed geographic information and maps. The USGS launched an ambitious program to meet their needs. Four parties were fielded

In 1896, Josiah Edward Spurr led a party of Survey geologists over Chilkoot Pass to the Yukon River and downriver to St. Michael. Shown here, left to right, are Frank C. Schrader (1860-1944), Spurr (1870-1950), and Harold B. Goodrich. The party stopped for special investigations in the Fortymile, Birch Creek, Eagle and Rampart gold districts, and was in the region at the time of the Fortymile and Klondike gold strikes. In 1898, Spurr made a 1,400-mile reconnaissance in southwestern Alaska, from Tyonek across the Alaska Range to the Kuskokwim drainage and downriver to Kuskokwim Bay, then back by way of rivers and lakes to the Alaska Peninsula and across the peninsula to Katmai. (U.S. Geological Survey)

in Alaska in 1898. The topography of the Fortymile River gold district was surveyed by one party. Another surveyed topography and geology along the White and Tanana rivers. The latter group included a young geologist new to Alaska but destined to leave a mark here: Alfred Hulse Brooks.

Brooks had graduated in 1894 with a geology degree from Harvard, where he was a contemporary of Josiah Spurr. After graduation, Brooks landed a USGS summer job, assisting Willard Hayes in the southern Appalachians. He soon fell under the spell of the Alaska veteran's tales. In 1898, after attending a geologic conference in Russia and touring parts of Europe, Brooks was in Paris, studying at the Sorbonne, when he received a telegram offering him a Survey job in Alaska. He wired his acceptance and departed Paris within a day to embark on what became a long and illustrious career in Alaska geology.

Besides the two parties in the Yukon-Tanana Uplands in 1898, two Survey field parties penetrated the Alaska Range starting from Tyonek on Cook Inlet. One party surveyed the geology and topography of the northeast Susitna drainage, around Broad Pass. Josiah Spurr and topographer W.S. Post led another party up the Yentna and Skwentna rivers, then portaged their cedar canoes to a tributary of the upper Kuskokwim River, which they descended to Bethel. They were only half done. From the shore of Kuskokwim Bay, they ascended a river and portaged to Togiak Lake. They crossed mountains and lakes to Nushagak, on Bristol Bay, then to Naknek, and via Naknek Lake, across the Alaska Peninsula to Katmai. The hardy explorers had traversed 1,400 miles.

Spurr's report on the journey covered topography, geology and mineral resources, with notes on Natives encountered, fish and game, timber, birds and animals. All was delivered in the intrepid geologist's rather literary style. Two years after the exploit, Spurr was honored by the naming of Mount Spurr volcano, 70 miles west of Anchorage.

Walter C. Mendenhall (1871-1957) was a USGS geologist for 49 years, and was USGS director from 1930 to 1943. This photo was taken around 1931. In 1898, attached to an Army expedition under Capt. E.F. Glenn, Mendenhall explored from Seward to Knik, up the Matanuska Valley and across the Alaska Range to the Tanana River. In 1901, he explored and mapped from Fort Hamlin on the Yukon River to Kotzebue Sound. He also worked on the Seward Peninsula and in the Copper River area. (U.S. Geological Survey)

Two other Survey geologists were in the field in Alaska that year, attached to Army expeditions. Frank Schrader, who had assisted Spurr in the upper Yukon two years before, worked in Prince William Sound and the lower Copper River valley. Walter C. Mendenhall surveyed geology and topography along the trail from Resurrection Bay to Turnagain Arm and from Crow Creek to Eagle River. From Knik village, Mendenhall traveled up the Matanuska to the Tanana River, noting the Matanuska Valley's coal deposits.

Walter Mendenhall achieved distinction in subsequent years, but it was not for him that Juneau's famous glacier was named. That honor fell to Thomas C. Mendenhall, superintendent of the U.S. Coast and Geodetic Survey from 1889 to 1894. Walter Mendenhall was the fifth USGS director. From 1930 to 1943, through the Depression and into World War II, he successfully championed the cause of basic science, proclaiming, "There can be no applied science unless there is science to apply."

Before the close of the 19th century, William Dall put in one more appearance in Alaska. In 1899, Dall participated in the celebrated Harriman Alaska Expedition, which began in part as a big game hunt and summer holiday for railroad baron Edward H. Harriman and his family. It grew into an important, if unusually luxurious, natural history expedition. With the help of the Smithsonian Institution and the Washington Academy of Sciences, some 30 men of science were recruited from public and private organizations to join the expedition. Dall was one of three senior USGS scientists on the cruise.

A 250-foot steamship, the *George W. Elder*, was refurbished for Harriman's cruise.

Gold rushes, for a time seemingly one right after another, prompted government officials in Washington D.C. to re-evaluate the U.S. Geological Survey's budget. Although it took Congress a while to comply with the administration's request for a substantial increase, it did eventually come through. The rush to Sunrise on the Kenai Peninsula, shown here in 1904, caught the Survey's attention in the 1890s when, for a brief moment in summer 1898, the town was reportedly the largest community in Alaska. (F.H. Moffet Photo No. 196, U.S. Geological Survey)

Among the passengers, there numbered five botanists, three zoologists, three ornithologists, a general biologist, a forester and an anatomist. There were two geologists, a paleontologist, a mineralogist and a mining engineer. John Muir was aboard and lectured on glaciers, ornithologists discussed birds, Dall held forth on Alaska history and geography and even offered a poem, "The Song of the Inuit." The ship logged 9,000 miles with about 50 stops between July 1 and Aug. 31.

After the voyage, some 50 specialists worked on the huge collections that had been gathered. The fruits of the expedition were 12 lavishly illustrated volumes, jointly published by Harriman and the Smithsonian Institution.

As the century turned, Congress and the public clamored for detailed maps and geologic investigations of mining districts. In 1899, Alfred Brooks had postponed sailing home from St. Michael to follow goldseekers rushing to Nome. Starting in October, Brooks and two colleagues had mapped on foot a 10-mile radius around Nome. Typhoid fever was endemic in the unsanitary boomtown and Brooks contracted the disease. He was critically ill for weeks after he returned to Washington.

In 1900, the Survey dispatched eight topographers and six geologists to mining districts in Prince William Sound and the Copper River basin, as well as the Seward Peninsula. While Brooks returned to the gold districts around Nome, Walter Mendenhall mapped the eastern Seward Peninsula. The two parties mapped the topography and geology of more than 6,000 square miles, surveying more than 100 creeks in nine gold districts.

In 1901, Brooks continued his work on the Seward Peninsula, but he also spent time in mining districts around Ketchikan and elsewhere in Southeast. Mendenhall led a party from Fort Hamlin on the Yukon River to Kotzebue Sound, via the Dall, Kanuti, Alatna and Kobuk rivers. Topographer W.J. Peters and geologist Frank Schrader traversed the Arctic Mountains, as the Brooks Range was then known, providing the first reliable information about that great range. From the Yukon River they crossed through Anaktuvuk Pass and via the Colville River to the Arctic Ocean. They followed the coast to Point Barrow and on to Cape Lisburne.

The following year, Brooks led an arduous trek. Seven men and 20 packhorses departed Tyonek and crossed the Alaska Range through Rainy Pass. They skirted the northwest base of the range for 200 miles to

In 1902, Alfred Hulse Brooks led seven men and 20 packhorses on an 800-mile geologic and topographic survey of the unexplored northern slopes of the central Alaska Range. The party consisted of Brooks, topographer D.L. Reaburn, assistant geologist L.M. Prindle, Odell Reaburn as topographic recorder and three camp hands. They departed Tyonek on June 2. The first problem they met was the swollen Skwentna River, which foiled repeated attempts to ford it. Finally crossing with the aid of a boat, the explorers plunged into a region of heavy brush and timber, which required continual trail chopping. On July 15, they discovered a pass, which they called Rainy Pass, and crossed to the Kuskokwim drainage. Travel was easier on the northern side of the range. The main obstacles to progress were the many swift, braided streams. Horses were occasionally swept off their feet and rolled over, though none were lost to streams. Moving camp an average of eight miles a day, Brooks and Prindle mapped the rocks and geologic structure exposed to view, while Reaburn surveyed the topography in a corridor roughly 10 miles wide along their route. On Aug. 3, camp was made some 14 miles from Mount McKinley, which had never before been approached by non-Natives. Brooks took a day for a side trip to a ridge within nine miles of the summit. Based on his observations and Reaburn's topography, the pair later wrote "Plan for Climbing Mount McKinley," an article for National Geographic *(1903). Pressing eastward across the Nenana, the party met some Natives and a white man on Aug. 24, the first people they had seen in nearly three months. On Sept. 1, they crossed the Tanana, headed for Rampart, on the Yukon. "To the natives" on the Tanana, Brooks wrote, "the arrival of white men from the mountains seemed little short of miraculous." The Tanana Natives told the explorers the lowland of the Tolovana River, between them and the Yukon, was impassable for horses. And the route did prove arduous. In six days, they built five bridges and four rafts to cross streams. The lowland passage was hellish for the horses, and the frost-bitten grass on the uplands the rest of the way offered scant nourishment for the weakened beasts. Only 11 horses survived until the party reached Rampart, on Sept. 15, 105 days after leaving Tyonek. (U.S. Geological Survey)*

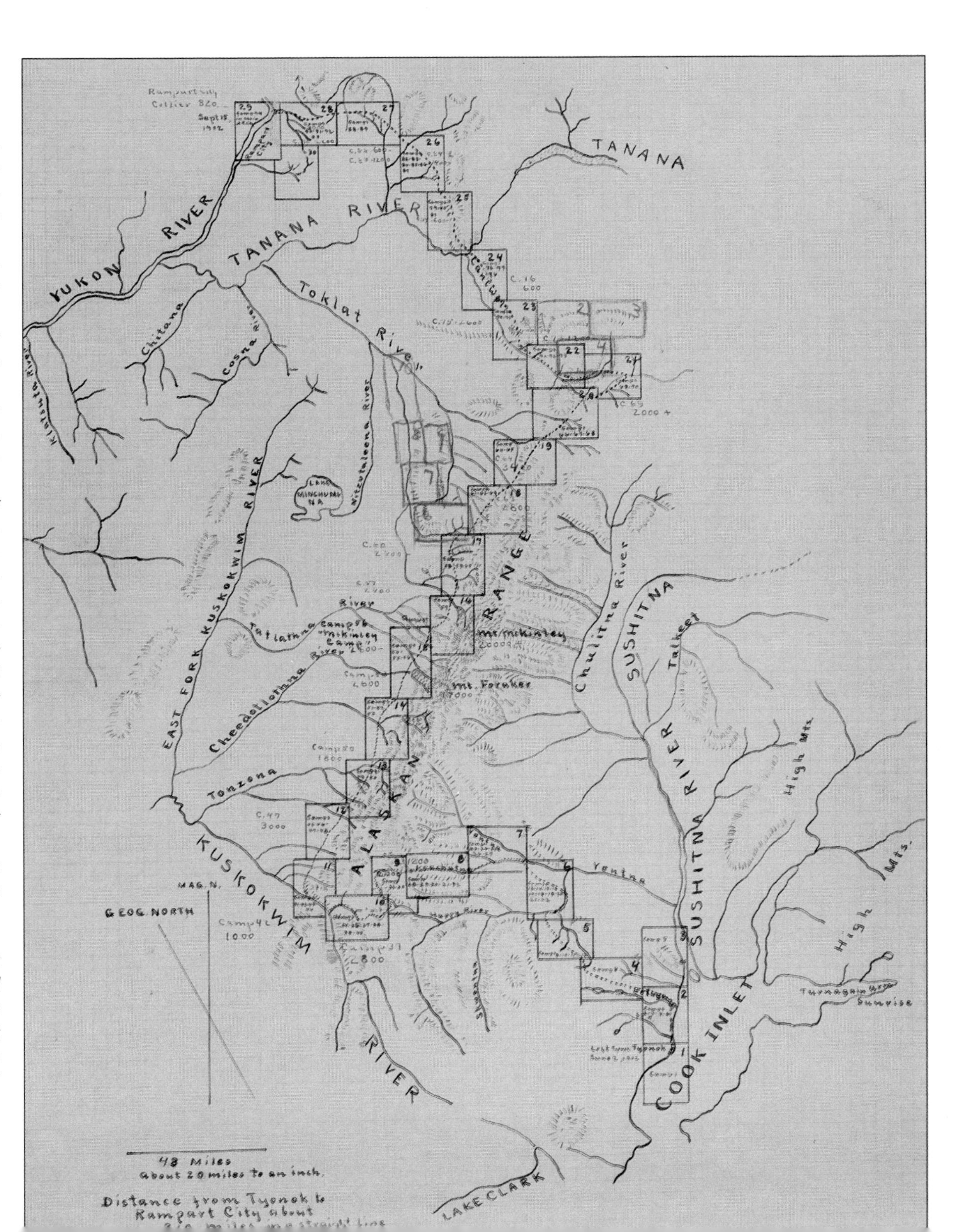

the Kantishna River drainage. Passing Mount McKinley, to ascertain what he could of the great mountain's geologic structure, Brooks climbed as high as snow and ice would allow. From the Kantishna area, the party struck northwest across the Tanana and Tolovana rivers to Rampart on the Yukon. They arrived after 105 days, having surveyed topography and geology through 800 miles of tough terrain.

Alfred Hulse Brooks (1871-1924) left a legacy in Alaska geological studies as wide as the mountain range named for him. He first came to Alaska in 1898 and, after pioneering expeditions, was appointed chief Alaskan geologist. (Historical Photograph Collection, photo no. 83-209-134N, Archives, University of Alaska Fairbanks)

The Brooks Years: Assailing the Flanks of Mountains

Were the day clear, I could see Mount McKinley from the window. As I picture in my mind its stupendous height, I compare it to our science. Many have assailed its flanks; some have proclaimed untruths about it; some have climbed by great effort well up the slopes; a very few, the best by natural selection, have reached the summit and there attained the broad vision denied those at lower altitudes. As for me, I am satisfied to have been able to traverse the great lowland to the base and to climb the foothills.

—*Alfred Hulse Brooks*
quoted by P.S. Smith,
in Memorial to Alfred Hulse Brooks,
Bull. Geol. Soc. of Am., v 37, 1926

In 1903, with the USGS' Alaska program thriving, the agency established a separate Division of Alaskan Mineral Resources. Alfred Brooks headed the new division. He had been in charge of Alaska geology for two years and remained so for another 21 years, until his death. At one point, Brooks refused promotion to the position of Chief Geologist, preferring to remain at his Alaska post.

The new Division of Alaskan Mineral Resources was a microcosm of the larger Survey. It had its own topographers as well as geologists. The age of exploration in Alaska was far from over in 1903. As late as 1908, Brooks reported to the National Conservation Commission that 84 percent of Alaska was "geologically, almost unknown." Less than 1 percent had been mapped in detail.

Despite the reconnaissance work that remained, Brooks saw clearly beyond it to a time of economic development. And the development of Alaska's resources, he believed, depended on transportation. Roads were important, Brooks said, but economical, long-distance transport required railroads. The location of railroads should be rationally based on topography and the distribution of mineral resources, he argued. In 1904, Brooks asked for money to study possible railroad routes. He was slightly ahead of his time.

Gold, silver and copper held the greatest allure for Alaska miners in the early 1900s. But Brooks felt that energy resources — coal and oil — were equally vital to future development. He hired George C. Martin, then with the Maryland Geological Survey, to study oil in Controller Bay and coal along the Bering River, both near Katalla.

Oil seeps near Katalla were common, having first been described in an 1898 Survey report. A private company started an exploratory well in Controller Bay in 1901 but stopped when drilling tools were lost in the hole. In 1902, the company drilled a second hole, George Martin reported. "At a depth of 360 feet the tools appeared to break through into a cavity of the rock and a large flow of oil began, spouting, it is reported, many feet above the top of the derrick."

Two other companies began drilling wells near Katalla the next year. In September 1904, 15 wells had been drilled or were in progress, according to Martin. One produced petroleum for fuel at other wells, but none was producing oil in commercial quantities. Interest in the Controller Bay oil fields ebbed with the discovery of oil in California, where production costs were lower. Survey interest in fossil fuels continued, however, as studies focused on coal and oil in Cook Inlet and coal in the Matanuska Valley and on Cape Lisburne.

By 1911, Alaska development was a subject of national debate. Secretary of the Interior Walter L. Fisher toured Alaska in late summer, with Alfred Brooks. Returning to Washington, Fisher embraced Brooks' recommendation for railroad construction. He also recommended self-government for Alaska.

President Taft, in his State of the Union speech and a special message to the Interior Department, endorsed many of Secretary Fisher's proposals. He favored federal construction of Alaska railways but failed to support self-government. In April 1912, the U.S. House of Representatives went beyond Taft's position and passed a bill to create a legislature in Alaska. The Senate passed the bill in July and a month later, Taft signed it. The Territory of Alaska was born.

Section 18 of the territorial act established a federal railroad commission to study transportation questions and examine possible railroad routes from tidewater to interior coal fields and navigable waters. When the commission was organized, Alfred Brooks was named vice chairman.

If 1912 was a year of momentous events for Alaska, one geologic event fit the theme. On June 6, the greatest recorded volcanic eruption of the 20th century ripped the Alaska Peninsula near Katmai. Kodiak, 100 miles away, was plunged into night by a foot of ashfall; ash-laden boats foundered in the harbor. The darkness lasted more than two days. The blast was heard 140 miles away, and ash dusted Ketchikan and Seattle. It was later found that ash from Katmai, injected into the stratosphere, lowered the average temperature in the Northern Hemisphere by 2 degrees F for more than a year.

BELOW: *George Curtis Martin, 1875-1943, spent 17 years in Alaska as a USGS geologist between 1903 and 1924. Martin's first project was to assess the Katalla oil field and the nearby Bering River coal deposits. In 1912, with funding from the National Geographic Society, Martin was the first geologist to study the Katmai volcanic eruption. (U.S. Geological Survey)*

RIGHT: *As miners rushed from one reported bonanza to another, the press wasn't far behind. Two years after a gold find in the Tanana Valley gave birth to Fairbanks, the* Fairbanks News *had their own collection of gold field tales. (Early Prints, 01-4203, Alaska State Library)*

Congressional budget-cutting and delays had left the USGS with no money for Alaska field work in 1912, but George Martin studied the Katmai cataclysm anyway. The National Geographic Society funded his trip. Martin arrived four weeks after the great explosion and spent a month documenting the eruption's effects. Then he lingered to study mineral deposits on Kodiak Island.

In 1917, topographer James W. Bagley produced a report which helped revolutionize the way the USGS made maps. In an 88-page monograph, *The Use of the Panoramic Camera in Topographic Surveying* (USGS Bulletin 657), Bagley summarized pioneering work he had done in Alaska. The standard method of topographic mapping employed a plane table, a drawing board mounted on

a tripod, fitted with a ruler equipped with a sight. The topographer notes angles to landmarks, calculates distances by triangulation and sketches in contours of equal elevation. But in complex terrain, according to Bagley, "an almost bewildering number of peaks, spurs, pinnacles, saddles, and slopes" can make plane table mapping tedious, difficult and slow.

An alternative method, phototopographic mapping, employs photographs from which topographic contours are mapped using optical instruments in the office. The method, developed in France, was slow to gain acceptance in the United States. It was first used in Alaska in 1904, by two brothers who worked for the USGS. Their innovation was in using a panoramic camera, with a specially wide field of view, but little more was done with their camera until 1910, when Bagley began experiments in Alaska. He used both a panoramic camera and plane table and started slowly, but in 1913, Bagley mapped 2,500 square miles in Broad Pass.

Bagley's pioneering phototopographic techniques were soon adopted by topographers outside Alaska. World War I gave a major boost to mapping from aerial photographs, which Bagley had discussed in his monograph. Topographic mapping from aerial photographs is the standard method today.

World War I wrought other changes in USGS work in Alaska. Concern for strategic minerals added mercury, platinum, tungsten, tin, antimony, nickel, lead and zinc to the list of metals being sought.

Field work in Alaska during the war was typified by John Beaver Mertie's field season in 1917. The year the United States entered the war marked Mertie's seventh summer in Alaska. It began with a chartered boat trip from Ketchikan to Lituya Bay, where he was to investigate a reported nickel deposit. After narrowly avoiding disaster on the rocks of the bay's current-plagued entrance, Mertie found and studied the deposit. He also encountered a bear, collected fossils and panned the bay's beaches, turning up traces of platinum and gold.

Returning to Ketchikan, Mertie caught a steamship to Seward, where he visited area gold mining camps. Another steamer carried him to Seldovia, where he was to examine a deposit of chromite ore. The chromite cropped out on a nearby mountain, but the slope was impassable with devil's club. A Seldovia Native knew a trail to the site and agreed to show Mertie. They set out one morning at 2 and spent five hours climbing and chopping

Geologists used horse-drawn sleds to haul their gear through Thompson Pass out of Valdez in 1909. (Stephen R. Capps Collection, photo no. 83-149-954, Archives, University of Alaska Fairbanks)

the overgrown trail. Atop the mountain, Mertie mapped and sampled the chromite veins. Then they descended, regaining the foot of the trail around midnight.

Mertie's last task that summer was to study the Cache Creek gold district, south of Denali. He secured passage from Anchorage as an unpaid hand aboard a riverboat bound up the Susitna and Yentna rivers to a Kahiltna River camp. From there, Mertie hitched a ride to the Cache Creek mining camps. His conveyance was an old truck drawn by horses over a primitive wagon trail. Part way there, he was disconcerted to learn that the jouncing truck's cargo was dynamite.

Mertie spent many days traveling on foot along rain-swollen creeks from mine to mine. He had several close calls fording the creeks and was wet and bone-cold much of the time. He talked with the miners, made notes and panned creeks himself, turning up traces of platinum. He collected some fossils and then headed home, to study his samples and report on the summer's work.

While war spurred the quest for strategic minerals, construction of the Alaska Railroad began to open the railbelt. From Seward to Fairbanks, there was demand for topographic maps and information about mineral resources, especially gold and coal. To support this work, the USGS opened its first office in Alaska, in 1918, in the railroad boomtown of Anchorage.

Interest in northern Alaska also began to pick up after the war. In 1919, the USGS published a 251-page report on Alaska's eastern Arctic coast. The report was by Ernest de Koven Leffingwell, a private geologist who had lived among the Inuit on Flaxman Island, some 40 miles east of Prudhoe Bay, from 1907 through 1914.

John Beaver Mertie Jr. (1888-1980) spent more than 30 field seasons in Alaska. Mertie is shown here at age 27, in 1915, the year of his first independent work as a USGS geologist. Mertie and topographer Rufus Sargent (1875-1951) worked with a horse pack train in southwestern Alaska that year. Mertie reported that miners in Iditarod were so desperate for horses that when Sargent auctioned theirs at the end of the season, they netted $1,200 in gold. (U.S. Geological Survey)

Working alone or with Inuit companions, Leffingwell compiled maps of the Arctic coast and studied geology from Point Barrow to the Canadian border and as far as 50 miles inland. His studies of permafrost were pioneering and he identified the Sadlerochit Formation, the rock unit which forms the main reservoir of the Prudhoe Bay oil field. His work, conducted at his own expense, was outstanding and the USGS was pleased to publish the results as Professional Paper 109. Leffingwell's driftwood-and-sod home on Flaxman Island is now a State Historic Site.

In 1923, a large portion of the west-central Arctic Slope, including oil seeps near Barrow, was designated Naval Petroleum Reserve 4. From 1923 through 1926, the U.S. Navy appropriated funds for USGS studies in NPR-4, as the 37,000-square-mile reserve was called. Philip S. Smith headed the effort. A former Harvard instructor, Smith had first worked in Alaska in 1906.

In January 1924, Smith, Mertie, two topographers and two assistants set out from USGS headquarters in Washington on an expedition to NPR-4. Rail, steamship and rail carried the explorers as far as Nenana. Dogsleds took them to Tanana, where a dog musher and a cook completed the crew. Smith's plan was to cross the mountains by dogsled in winter and be poised on a stream bank when spring reached the Arctic Slope. Then they would split into two parties of four, headed by Smith and Mertie, each to map topography and geology on their separate ways to the coast.

The group set out with four sleds, 54 dogs and four nested canoes to be used after breakup. Six additional sleds were hired to reach the southern foothills, plus 90 more dogs and five freight handlers who transported four tons of food and gear. After a month of hard travel, they attained Survey Pass. Another month was spent transporting

Members of the 1924 USGS expedition to the newly designated Naval Petroleum Reserve 4, on the west-central North Slope, assembled in Tanana before departure, in February. From left are: John A. Swenson (freight handler), Walter R. Blankenship (dog musher), Frank B. Dodge (recorder/note-taker), Richard K. Lynt (topographer), Mr. Goss (Northern Commercial Co. agent), Philip S. Smith (geologist in charge), John B. Mertie Jr. (geologist), Gerald FitzGerald (topographer), Harry A. Tait (recorder/note-taker), Titus Nicolai (freight handler). Arriving on the banks of Killik River in winter, the explorers, minus the freight handlers, split into two groups headed by Smith and Mertie that made separate ways to the arctic coast by canoe. Both parties gained Barrow on Aug. 25, within 12 hours of one another. (U.S. Geological Survey)

gear and setting up camp 20 miles north of the divide, on the Killik River.

Weeks passed awaiting breakup. Temperatures dropped at times to minus 60 degrees. The supply of dogfood dwindled, and the dogs grew thinner. Many had to be shot. Six dogs remained on May 31, when ice on the Killik began to move. The surviving dogs joyously leapt into the canoes as the explorers pushed off at last into the icy river.

The summer of 1924 held a series of woes for the explorers. The first day on the Killik, ice slashed holes in their canoes. A few days later, Mertie broke his foot. There were hazardous rapids, clouds of mosquitoes and unnerving uncertainty about which tributaries to follow and where to portage. On Aug. 25, Mertie's party reached Point Barrow, and Smith's party arrived 12 hours later. They soon learned that their ship to Nome had been crushed by sea ice and sunk. Fortunately, they found passage on a Bureau of Education ship which had avoided the ice. On Sept. 5, the scientists reached Nome.

On Nov. 22, 1924, Alfred Hulse Brooks collapsed at his desk in Washington, D.C., and died. The chief Alaska geologist was 53. Two years earlier, the USGS had reorganized and the Division of Alaskan Mineral Resources had metamorphosed into the Branch of Alaskan Geology. For 21 years, Brooks had headed first the division then the branch. Philip Smith succeeded Brooks and was chief Alaska geologist for the next 21 years. The year after his death, Brooks was honored by Smith, who renamed the Arctic Mountains, Alaska's rugged northern arc of alps, the Brooks Range, in lasting tribute to a giant in Alaska geology.

Philip Smith at the Helm: The USGS in Peace and War

Starting in the late 1920s and lasting through the Depression, budgets were tight for the USGS. Field work continued despite limited funds. Mertie and others mapped by canoe and on foot. Geologist Steven R. Capps, with various topographers, mapped in the southern Alaska Range with horses. One year he worked with a 17-horse pack train. In 1928, he flew with topographer Gerald FitzGerald and an assistant to a field site 80 minutes west of Anchorage by air. A horse pack train that carried support equipment to the site took 20 days to make the journey. Budget analysts figured the savings in food and salaries for the scientists just about offset the cost of the flight.

At about this time, aerial photography was first used for mapping in Alaska. An experimental mapping project in the Ketchikan mining district was carried out with Navy aid.

In the early 1930s, the Alaska Railroad was losing money and Congress wanted a way to boost tonnage on the federal railway. As a spur to resource development, the USGS was funded to study mineral and coal deposits in the railbelt. Stephen Capps led the effort, from 1931 through 1934. Among other things, he visited mining camps along the route, gleaned information about the ores being mined and compiled production statistics. He returned to the area in 1936.

Basic research was not wholly neglected, even during the Depression. Philip Smith took time to study the route of early man from Asia to North America. At a 1937 science congress, he presented the geologic and geographic evidence for a migration route across the Bering Strait, rather than up the Aleutians, a route then favored by some anthropologists.

As the storm clouds of war gathered once more, concern over strategic minerals deepened and the lean years of the Depression gave way to a flurry of geologic investigations. There were 17 strategic minerals projects in Alaska in 1941, seeking chromium, nickel, tin and mercury, among other metals. Seventeen geologists were hired in 1942, to staff 29 more minerals projects, including studies targeting coal, oil and a new strategic metal: uranium.

Also vital to the war effort was engineering geology: studies of surface materials, including sand and gravel needed for construction. Engineering geology also figured in the location and design of roads, airfields and other installations. Field surveying for topographic maps almost ceased during World War II, supplanted for the time being by more rapid mapping from aerial photographs. Late in the war, and for some years afterward, mapping and studies of permafrost were undertaken for the military. Radioactive minerals were sought for the Atomic Energy Commission. In 1945, a study of Aleutian volcanoes was started, to assess hazards posed to military installations and aviation by volcanic eruptions. Robert R. Coats and others ranged from island to island, compiling maps and information on Aleutian volcanoes.

A modern program of petroleum exploration was launched by the Navy in Naval Petroleum Reserve 4, in 1944. Between 1945 and 1952, a USGS office in Fairbanks oversaw drilling of 81 boreholes totaling some 170,000 feet in length or more than 32 miles. Nearly 3,750 miles of seismic surveys were completed within and adjacent to NPR-4. Extensive air photo coverage was obtained to aid mapping and early airborne magnetometers were used to map the Earth's magnetic field. Field geologists conducted studies first by foot and riverboat, then by a tracked vehicle called a Weasel, finally by helicopter. Begun as a wartime measure, the NPR-4 effort spanned a time of technological change which brought Alaska field geology, in many respects, from its pioneering roots

In 1928 the biplane Anchorage No. 2 *prepares to depart Anchorage for an 80-mile flight to a site west of Chakachamna Lake in the first plane-supported field work in Alaska undertaken by the U.S. Geological Survey. According to records with the photo, Matt Nieminen was the pilot, Gerald FitzGerald the passenger, with a man named Cope ready to crank the propeller. Fitzgerald was the topographer for the Capps expedition. (Stephen R. Capps Collection, photo no. 83-149-2770, Archives, University of Alaska Fairbanks)*

into the modern age. In widespread use by geologists after 1952, helicopters in particular enabled geologists to map large areas far more rapidly, and to reach sites previously all but inaccessible.

Near the beginning of this time of transition, in early 1946, Philip Smith retired as chief Alaska geologist as the USGS reorganized for peacetime pursuits. Topographic mapping in Alaska was transferred to a Denver-based branch of the Topographic Division. In October 1946, the old Branch of Alaskan Geology was dissolved, some functions were transferred to other units of the Survey and a new Alaskan Section was constituted with responsibility for geologic mapping and mineral resource studies. Section offices opened in Juneau, in 1946, and in Fairbanks, in 1949. The post-war transition period ended in 1952, when the modern Branch of Alaskan Geology was born, in roughly its present form.

Modern-Day Pioneers

The years after World War II brought not just new technology and reorganization but social transformation as well. The first woman scientist who worked for the USGS in Alaska was Mary E. Hill, a member of an airborne magnetometer crew who spent two months surveying the magnetic field over NPR-4, in 1945. In 1949, geologists Florence Weber and Florence Collins joined the Navy Oil Unit in Fairbanks.

The Survey was not eager to send women geologists into the field on the North Slope in 1949. But Weber and Collins were both pilots, and on summer weekends, they flew Weber's float-equipped plane to lakes in the Kantishna area. They would camp and map the geology around the lake for a few days. In the late 1950s, Troy Pewé, then with the USGS, began to include Weber and Collins in official field parties. "He'd been going into the field with his wife for years, so it didn't bother him" to work with women, Weber recalls. Pewé later became a distinguished professor at the University of Arizona. Weber and Collins published their work. Collins later retired to raise children, but Weber stayed on to become one of the Survey's experts on Yukon-Tanana geology. She retired in 1990, after some 40 years of service, but she and her husband still live in Fairbanks and on most days, Weber goes into the office to work on her maps and manuscripts.

During and after the war, the USGS expanded its scope in Alaska beyond topographic mapping and minerals-oriented geology. Offices in Palmer and Juneau pursued studies of water resources for power and water supply. A public information office opened in Anchorage in 1953, to disseminate Survey maps, reports and information.

Studies of permafrost and engineering geology were expanded. So were seismic monitoring and earthquake investigations, and geochemical and geophysical surveys — studies of soil and water chemistry and of

Tracked vehicles called Weasels were used on the North Slope and northern foothills of the Brooks Range during mapping of Naval Petroleum Reserve 4 from 1945 to 1952. Two Weasels are shown fording the Oolamnagavik River, a tributary of the Colville River, in 1950. (U.S. Geological Survey)

The Kuskokwim River winds through a regional belt of gold and mercury deposits in southwestern Alaska. (Marti Miller, U.S. Geological Survey)

the Earth's magnetic and gravitational fields and radioactivity. In 1953, topographic mapping at a reconnaissance scale was completed: 153 quadrangles at a scale of 1:250,000 covered the territory. More detailed mapping, at a scale of 1:63,360 (or 1 inch represents 1 mile) is still underway.

Prior to the 1940s, most USGS geologists who worked in Alaska traveled to the field from Washington, D.C. each year. A few were based in Alaska for special purposes, a practice more prevalent after the war. In the 1950s, the Survey established a regional center in Menlo Park, Calif., south of San Francisco. Among the units transferred there from Washington, D.C., was the Branch of Alaskan Geology headquarters.

As the Cold War with the Soviet Union intensified, strategic minerals again became the focus of national concern. New technologies demanded new materials, including new alloys and radioactive elements. The USGS undertook a Heavy Metals Program to locate critical minerals. Much effort was spent in Alaska on the hunt for strategic minerals.

On March 27, 1964, Southcentral Alaska was rocked by the second largest earthquake ever recorded, exceeded only by a 1960 Chilean temblor. Moment magnitude is the index seismologists prefer for large earthquakes. It is a measure of the total energy released, and the Good Friday quake, centered beneath Prince William Sound, had a moment magnitude of 9.2, equal to 73,000 Hiroshima-size atomic bombs.

The quake jolted the Alaska geologic community along with everyone else associated with the state. Federal, state, university and even industry geologists all expended much energy in the temblor's aftermath. USGS scientists from California and elsewhere flew to Alaska to study the disaster. Faults and landslides were mapped and their movement was measured. Clay-rich deposits and soils susceptible to failure during quakes were studied and mapped. Scientists spread out across Prince William Sound to visit communities near the earthquake's epicenter, many of which were hardest hit not by the quake but by tsunamis, ocean waves created by the temblor.

Efforts to understand the Good Friday earthquake continued for many years. The USGS published a series of 28 Professional Papers on the quake, presenting geologic observations and documenting the temblor's effects on Alaska communities, water supplies and hydroelectric structures, electric utilities, railroads and highways. A number of these were written by George Plafker and others in the Branch of Alaskan Geology. Findings were also disseminated through conferences and in professional journals.

Four years after the quake, in 1968, a jolt of another sort shook Alaska's geologic community. Atlantic Richfield Corp. announced the discovery of a giant oil field near Prudhoe Bay. Subsequent construction of the trans-Alaska pipeline to carry the oil to market required a small army of engineering

The late Farid Kutyev of the Russian Academy of Sciences in Petropavlovsk inspects chromite seam at Red Mountain deposits near Seldovia. A past producer of chrome, the area has also been recently evaluated for industrial minerals such as olivine. (Thomas K. Bundtzen)

geologists, including Oscar Ferrians and others of the USGS. They mapped permafrost, soil and surface deposits along the pipeline route and the associated haul road.

Work by Arthur H. Lachenbruch, of the USGS, illuminated special problems in building a pipeline on permafrost. The friction of oil flowing through a pipe warms the oil, and the resulting heat can thaw permafrost and destabilize the pipeline's foundation. To accommodate concerns raised by Lachenbruch, the trans-Alaska pipeline was altered in route and design. It traversed gravel and bedrock where possible and was built above ground in permafrost terrain, on specially insulated support structures which dissipate heat into the air and not the ground. Roughly half the pipeline is above ground, the other half is buried.

The Prudhoe Bay oil strike, together with the 1974 Arab oil embargo, revitalized interest in the petroleum potential of NPR-4, west of Prudhoe Bay. The Navy renewed petroleum exploration in the reserve through a contract with a private company, Husky Oil. In 1976, NPR-4 was redesignated National Petroleum Reserve A, and management of the reserve was transferred from the Navy to the Department of the Interior.

Within Interior, the USGS was given responsibility for NPRA. The exploration program in the reserve was terminated in 1982, but by then, Husky Oil had drilled 28 test wells. Nearly all found indications of oil, natural gas or both, but none tapped deposits large enough to warrant development. Husky also drilled six wells in the existing natural gas field near Barrow, to supplement the gas supply in local use since the field's discovery in 1949. Finally, the Survey began environmental studies in NPRA, to address the need to clean up the effects of years of petroleum exploration in the fragile Arctic Slope ecology.

Science in the Public Interest

Alaska had been a state for just 10 years at the time of the Prudhoe Bay discovery. In 1971, three years later, Congress passed the Alaska Native Claims Settlement Act. Among other things, ANCSA spelled out a process by which federal lands would be transferred to the State of Alaska and Native groups. The process of land selection among the three parties was complex but among the factors of interest to all was the resource potential of the lands in question.

To gather the information needed to make informed land selections, the state and federal governments launched ambitious programs of geologic mapping and mineral resource assessment. Central to the federal effort was the Alaska Mineral Resource Assessment Program of the USGS. Begun in the mid-1970s, the AMRAP goal was to evaluate the resource potential of Alaska lands in chunks defined by 1:250,000-scale quadrangles. More than 100 AMRAP studies were planned. A few quads on the Arctic Slope were omitted because the area's bedrock is largely concealed beneath glacial and marine sediment. Others that encompass mostly water were either dropped or were merged for study purposes with neighboring quads.

AMRAP studies were distinguished from earlier Survey work by their multidisciplinary approach. Tom Miller was branch chief from 1980 to 1985, during the AMRAP heyday. "It was a mineral assessment program but defined in a pretty broad sense," says Miller. In addition to mapping bedrock geology, each project team would map younger, surface deposits, geochemistry, aeromagnetics. "It would all culminate in a resource assessment of the area."

Scientists working in a given quad often shared encampments and helicopters. AMRAP team geologists generally came from the Branch of Alaskan Geology. Geochemists and geophysicists came from Survey branches composed of experts in those subdisciplines. "We probably had 10 or 12 AMRAPS going at one time" in the early 1980s, Miller recalls. During the summer, "we were running as many as eight helicopters at one time."

A typical AMRAP study devoted three field seasons of up to two months to work in a given quad. During the fourth year, a shorter time might be spent on detailed studies of targeted areas. The effort was labor intensive and the Branch of Alaskan Geology grew to meet AMRAP demands. Miller hired a number of new geologists, most were young and many were women.

In 1980, at the start of Miller's term as branch chief, headquarters shifted from Menlo Park, Calif., to Anchorage. Some branch members remained in Menlo Park, but new hires were concentrated in Anchorage. For a time, branch members in Anchorage were scattered in offices around the city. Branch members in Anchorage are now housed together in two buildings on the campus of Alaska Pacific University.

In the late 1980s, the AMRAP engine began to lose steam. The program's success may have accelerated its decline. Based on the AMRAP model, a similar program was drawn up for mineral studies in the Lower 48. When the two programs merged into a National Mineral Resource Assessment Program, Alaska projects had to compete for funding with work elsewhere. Some of the autonomy over AMRAP studies enjoyed by the Alaskan branch chief was eroded. Finally, as federal budgets grew tighter, Congressional support for the expensive venture began to soften.

The Alaska Mineral Resource Assessment Program is alive, but it funds mostly office studies these days. Two AMRAP projects were fielded in 1994. The 1995 budget may not support any AMRAP field work. The decline in the pace of field work does not mean the job is done. Of the 100 or so AMRAP studies planned, roughly a third are complete.

Despite the hard times, the Branch of

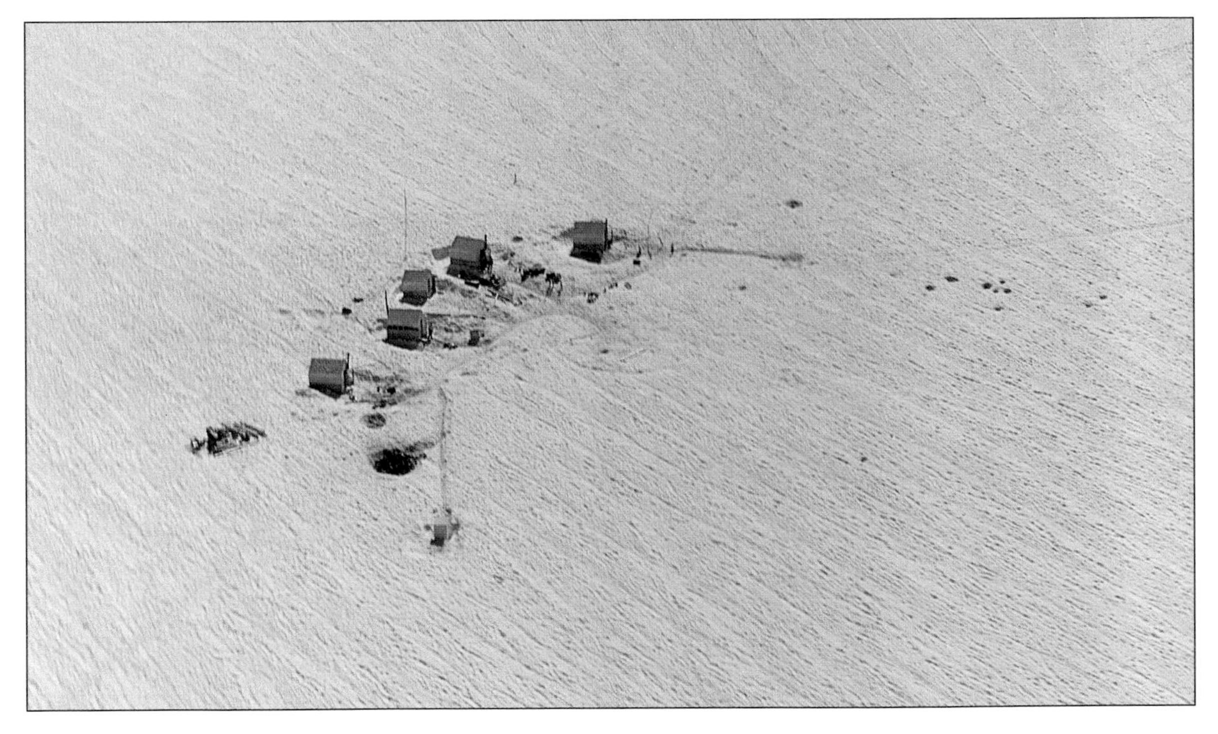

Brady Glacier, inside Glacier Bay National Park and Preserve, holds potentially large resources of cobalt, nickel and platinum. This 1959 aerial shows an ore sampling camp on the glacier. (Steve McCutcheon)

Alaskan Geology is busy. Since I.C. Russell tagged along with a Coast and Geodetic Survey crew 106 years ago, Survey scientists have always been resourceful in finding ways into the wilds. Marti Miller has arranged for Calista Corp. to pay her field expenses for work in an area that interests the Native group. Sue Karl has enjoyed support in the past from the U.S. Forest Service in Tongass National Forest, and may again. Alison Till has expertise in Denali National Park and has been asked to join National Park Service geologists in the field. Peter Heussler has cadged support for field work from colleagues at the University of Alaska in Fairbanks and the California-based USGS earthquake program. And so on. Work continues.

In recent years, Survey scientists working in Alaska have joined forces with the state of Alaska and the Russian Academy of Sciences in projects that compare geology and mineral resources of Alaska and the Russian Far East. Many branch scientists have visited Russian institutes and spent time in Russian field camps. Russian scientists have visited Alaska as well. Hopes are high the exchange will continue.

In April 1993, Dave Carter became chief of the Branch of Alaskan Geology. He is 13th in a line that traces itself back to Alfred Hulse Brooks. But the line will end with Carter.

USGS Director Gordon Eaton is reorganizing the agency and he plans to eliminate the branches. A series of "national programs" will replace them. Branch of Alaskan Geology scientists will probably be assigned to the mineral resources program. Layoffs are likely, a sober Carter says. "It's really uncertain what's going to happen with this branch."

The continued existence of the U.S. Geological Survey itself was in doubt after the 1994 national elections. Reports circulated that the Survey was among the agencies that would be closed to help balance the federal budget. The reports were premature. The agency survived Congressional budget hearings, although several years of deep cuts seem certain.

If the 116-year-old USGS has a future, a continued Alaska presence also seems assured. The Survey's Water Resources Division is relatively healthy, and studies of surface and ground water seem secure. Volcanic hazards programs are also sound. The Alaska Volcano Observatory comprises members of the USGS, the University of Alaska in Fairbanks and the Alaska Division of Geological and Geophysical Surveys. Observatory headquarters are in Anchorage, in a building shared with the Branch of Alaskan Geology.

Other Survey programs in Alaska face a murkier future. Members of the present Branch of Alaskan Geology will probably focus more narrowly on mineral resources. Other efforts may be cut back, including Dave Carter's own investigation of climate change in Alaska. Then again, there may be ways to muddle through until more bountiful days. Survey geologists in Alaska have always been flexible, ready to seize opportunities.

For more than 100 years, that has been the USGS tradition in Alaska. Witness what C. Willard Hayes wrung from Schwatka's expedition in Copper River country, in 1891. Or George C. Martin's work on the Katmai eruption of 1912, with National Geographic Society support. Picture John Beaver Mertie during World War I, riding a jouncing dynamite truck to the Cache Creek gold fields, or Florence Weber and Florence Collins flying Weber's plane to lakes of the Kantishna area. Scientists swarming across Southcentral Alaska after the 1964 quake. Or Marti Miller, John Gray and Elizabeth Bailey, engaged in an AMRAP study in the Sleetmute quad, seizing the chance to size up the implications of environmental mercury. Other examples are legion, and a complete listing would be a roll call perhaps hundreds of names long.

The devotion of these scientists and their colleagues has sometimes exacted a high human cost. Twenty-four USGS workers have died in the field in Alaska in the last 30 years alone, mostly by drowning.

The legacy of the USGS in Alaska would surely make Alfred Brooks proud. Through 1994, Survey scientists produced some 7,000 formal reports on Alaska. Thousands of articles in professional journals have also been written, countless talks presented at conferences. From 1978 through 1994, nearly 2,500 such articles and talks were penned or spoken, more than 150 per year.

The flood of information has contributed to the development and settling of Alaska, from Brooks' impact on the Alaska Railroad, to Lachenbruch's shaping of the trans-Alaska pipeline. Airlines depend on warnings of ash plumes from volcanic eruptions, urban planners rely on earthquake data. State agencies and Native groups use resource assessments to aid in land selections. Individual miners pore over Survey reports, topographic maps guide hikers.

If the shape of the Survey's future in Alaska is clouded, what shines through the politics of the day — what cannot be obscured — is the legacy of a century of toil, sweat and blood expended in Alaska by the men and women of the U.S. Geological Survey. ◗

Gold Production in Alaska by District, 1880 - 1993

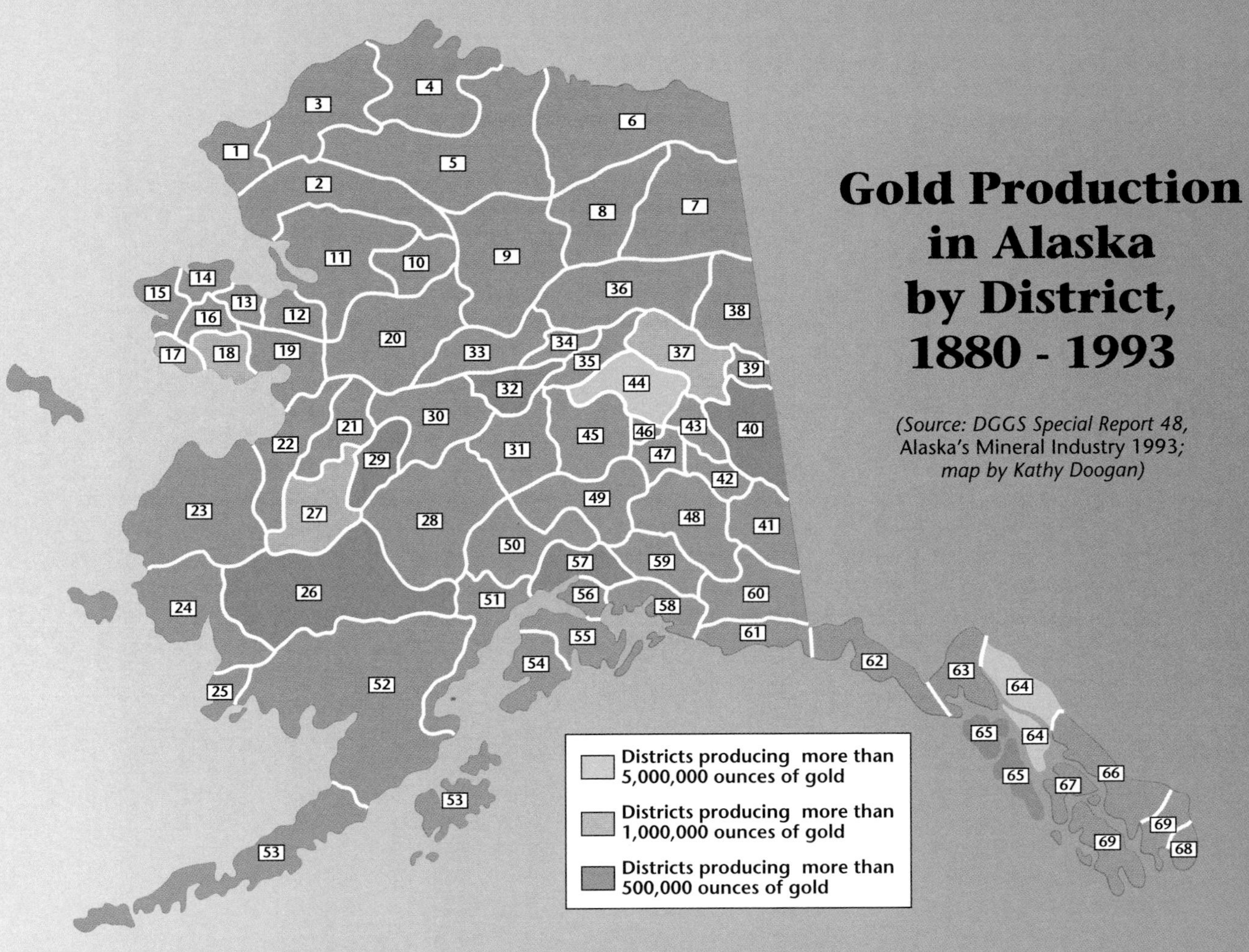

(Source: DGGS Special Report 48, Alaska's Mineral Industry 1993; map by Kathy Doogan)

Mining districts [A]	Production (in troy ounces)
1. Lisburne district	0
2. Noatak district	7,800
3. Wainwright district	0
4. Barrow district	0
5. Colville district	0
6. Canning district	0
7. Sheenjek district	0
8. Chandalar district	63,158
9. Koyukuk-Nolan district	324,804
10. Shungnak district	15,000
11. Squirrel River district	40,600
12. Fairhaven-Inmachuk district	347,671
13. Candle district	248,130
14. Serpentine district	4,220
15. Port Clarence district	40,946
16. Kougarok district	171,015
17. Cape Nome district	4,822,569
18. Council and Solomon districts	1,044,013
19. Koyuk district	83,532
20. Koyukuk-Hughes district	227,788
21. Kaiyuh district	5,400
22. Anvik district [B]	0
23. Marshall district	123,319
24. Bethel district	42,945
25. Goodnews Bay district	29,700
26. Aniak-Tuluksak district [C]	566,751
27. Iditarod district	1,559,260
28. McGrath-McKinley district	190,500
29. Innoko-Tolstoi district	696,047
30. Ruby-Poorman district	469,795
31. Kantishna district	99,307
32. Hot Springs district	560,532
33. Gold Hill-Melozitna district [D]	11,059
34. Rampart district	187,061
35. Tolovana-Livengood district	480,013
36. Yukon Flats district	0
37. Circle district	1,019,843
38. Black district	0
39. Eagle district	52,000
40. Fortymile district	530,265
41. Chisana and Nabesna districts	144,500
42. Tok district	0
43. Goodpaster district	2,350
44. Fairbanks district	8,275,576
45. Bonnifield district	80,492
46. Richardson district	118,570
47. Delta River district	5,647
48. Chistochina district	181,261
49. Valdez Creek district	427,875
50. Yentna-Cache Creek district	196,754
51. Redoubt district	105
52. Iliamna-Bristol Bay district	1,570
53. Kodiak and Unga Island district	112,400
54. Homer district	16

Mining districts [A]	Production (in troy ounces)
55. Hope-Sunrise-Seward district	130,178
56. Anchorage district [E]	0
57. Willow Creek-Hatcher Pass district	664,788
58. Prince William Sound district	137,715
59. Nelchina district	13,860
60. Nizina district	148,500
61. Yakataga district	18,040
62. Yakutat district [F]	13,200
63. Porcupine district	80,373
64. Juneau and Admiralty districts	7,260,248
65. Chichagof district	770,000
66. Petersburg-Sumdum district	15,000
67. Kupreanof district	0

Mining districts [A]	Production (in troy ounces)
68. Hyder district	219
69. Ketchikan district	62,000
Subtotal	**32,926,280**
Undistributed [G]	127,886
Total Production *(troy ounces)*	***33,054,166***
(metric tonnes)	***1,028***

[A] Mining district names and boundaries revised slightly from those defined by Ransome and Kerns (1954) and Cobb (1973). Sources of data: U.S. Geological Survey, U.S. Bureau of Mines, and Territorial Department of Mines Records 1880-1930; U.S. Mint Records 1930-1969; State of Alaska production records 1970-1993. Entries of "0" generally mean no specific records are available.

[B] Included in Marshall district.

[C] Includes Georgetown and Donlin districts.

[D] Includes Tanana precinct.

[E] Placer gold included in Willlow Creek/Hatcher Pass district.

[F] Includes lode production from Glacier Bay and placer production from Lituya Bay district.

[G] Production that cannot be credited to individual districts due to lack of specific records or for reasons of confidentiality.

Alaska's Mineral Industry Today

By L.J. Campbell

Editor's note: *Entries on some minerals contain information from* The Concise Columbia Encyclopedia, *3rd Edition, Copyright 1994 by Columbia University Press. Reprinted with permission.*

Much of the way Alaska is today can be traced to its minerals, those discovered and those still hidden. The search for them opened the land to settlements, some still thriving, others nothing but ruins in the grass. Humans took on audacious challenges against nature to reach these minerals, bridging rushing rivers, blasting rock walls, traversing glaciers, tunneling through frozen ground, trying the untried in a gamble for success. Miners today look at what has been in Alaska and what is to come, and they are a political force in the state's future. To understand something of Alaska today is to look at its mineral wealth and the people who mine it.

Minerals, Rocks And Ore

Alaska holds tons of gems and minerals: valuable metallic metals like gold, silver, zinc, copper and platinum, as well as precious and semiprecious gems, like diamonds, sapphires, garnets, jade, jasper and agate. Alaska has industrial minerals too, mainly gravel and construction stone but also marble, industrial garnets, zeolites, graphite and limestone.

Some of Alaska's minerals, essential to manufacturing, are economically important and sold to world markets. Zinc, for instance, is used in galvanizing iron and preparing alloys such as brass and bronze. The Red Dog Mine in northwestern Alaska is a major contributor to the world's supply of zinc. A significant percentage of automobiles currently manufactured in Japan, for example, contain zinc from Red Dog.

Mining opened Alaska to settlement and development, and historic trails like this one between Circle and Fairbanks make the country more accessible to today's hunters and outdoor enthusiasts. (Steve Seiller)

Other minerals are valued as gems and sought for jewelry and ornamentation, because of their beauty, clarity and durability. Poorer-quality gems are often used in industrial applications. Diamonds, among the hardest substances known, are used for abrasives and cutting. Diamond drills are a common tool in mineral exploration, because they can pierce the hardest rock.

Rocks are made up of minerals, usually combinations of many different minerals. An ordinary rock will hold minute amounts of more than a dozen metallic mineral elements, including gold, silver, platinum, mercury, copper, cobalt, nickel, chromium, lead, zinc, molybdenum, tin and tungsten.

Sometimes rocks hold great concentrations of minerals, 1,000 to 10,000 times amounts ordinarily found. These unusual deposits of minerals, or ore deposits, spark miners' dreams.

Ore deposits can be found on the surface, although most of the richest surface deposits have already been discovered. Yet there are still unknown ore deposits hidden in the ground. Geologists use a variety of methods to detect ore deposits. They look for telltale rocks known to occur in association with certain ores. They use computer programs that describe physical characteristics and chemical composition of each type of ore deposit and its relationship to the geologic environment, similar to the way biologists describe how an organism fits into a particular environmental niche.

Ore deposits take many forms. In a particular gold deposit off the Tok Highway in eastern Alaska, rhodonite holds veins of gold; in another deposit near Big Delta,

Types of Mining

Placer mining has historically been the most common form of metal mining in Alaska. Placer deposits are chunks and pieces of mineral mixed into gravels of existing or former stream beds. When found in former stream beds, the placers are called bench deposits.

Placer deposits sometimes formed as glaciers moved over hardrock formations, grinding them down and carrying off the minerals. Placers may be found on the surface or buried several hundred feet deep.

Gold commonly occurs in placer deposits, which attracted most of Alaska's early prospectors. But other minerals may occur as placers too. Platinum placers were mined at Goodnews Bay on the Bering Sea coast from the mid-1920s through the mid-1970s.

Placers may be mined at relatively low costs, making them attractive to individuals and small companies. At its simplest, placer mining requires nothing more elaborate than a shovel and gold pan; more typically, mechanical dredges dig the deposits, and a sluicebox is used to wash out the pay. Huge dredges operated around Nome and in the Fairbanks area for much of this century. Some placer miners use high-pressure hydraulic nozzles to shoot jets of water into rocky, silty banks to dislodge deposits. Several gold mining companies in Alaska do underground drift mining to reach buried placers; most use surface mining methods.

LEFT: *High-pressure nozzles shoot streams of water into placer bench deposits at a mine near Livengood, in the Interior. (Steven Seiller)*

ABOVE: *A placer miner on Dixon Creek, northeast of Nome, samples a hole for gold-bearing gravel. (Harry M. Walker)*

Surface mining, or open pit, methods include quarrying and strip mining. The soil and barren rock are removed to expose the ore, then the pit is excavated in a series of levels, or benches, arranged as a spiral or connected by ramps. The exposed ore may need to be broken apart using explosives. Then it's loaded and hauled away for processing.

Lode mining, in contrast, deals with mineralized hardrock. Lodes close to the surface may be mined by open pit methods. Otherwise, underground mining methods are used. Networks of tunnels may be built to reach the lodes, and the blasting, loading and hauling equipment is designed to work in small spaces.

Fort Knox will be the state's biggest producing hardrock gold mine when it starts up in 1996. The Fort Knox lode is almost pure gold in granite, and will be extracted using open pit methods.

Hardrock gold mining played big in Alaska's past, at Independence Mine in Hatcher

Pass of the Talkeetna Mountains; at the Nabesna Mine in the northern Wrangell Mountains; at the Apollo Mine on Unga Island in the Shumagins; at the Cliff Mine near Valdez; and at the Big Hurrah Mine near Solomon.

Most hardrock minerals must undergo processing at the mine site to concentrate the ore, or separate the valuable minerals from the host rock. Extractive metallurgy to concentrate the ore involves crushing the rock into small fragments; grinding it through ball or rod mills into fine powder; screening the ore to separate particles by size, so the coarser materials receive additional grinding; then extracting the minerals into high-grade concentrates. This extraction can be done through gravity or flotation in water, where the heavier minerals fall to the bottom or the lighter minerals rise to the top; or through hydrometallurgy, in which chemicals like cyanide, sulfuric acid or solvents are used to separate the minerals. The final stage is to remove additives and dry the concentrate to reduce its weight, then ship it to refineries and smelters.

Sometimes, a deposit will be *polymetallic,* which means it contains several valuable metal minerals. During the milling, the ore undergoes complex extraction processes to separate the various minerals from each other. For instance, in a massive sulfide deposit composed of iron, copper and zinc sulfides with traces of gold and silver, each of the valuable minerals must be separated so the concentrates of each are as pure as possible. Because copper and zinc require different smelting techniques, a copper smelter assesses a penalty when buying concentrates containing zinc; likewise, a zinc refinery will not pay full value for concentrates containing copper.

Two big Alaska polymetallic mines are Red Dog near Kotzebue, where zinc is the primary mineral with lead secondary, and Greens Creek on Admiralty Island, where silver was the targeted mineral during its first five years of operation. Green's Creek also produced significant amounts of gold, lead and zinc from its massive sulfide deposit.

—Parts adapted from *Alaska's Oil/Gas & Minerals Industry*

BELOW: *This aerial shows the "footprint" of the Kensington hardrock gold project 45 miles north of Juneau in 1990. Shown are the portal (entrance to the underground workings), waste rock and water treatment facilities. (Thomas K. Bundtzen)*

RIGHT: *Cambior Alaska Inc. mined placer gold deposits along Valdez Creek, south of the Alaska Range, for 11 years using open pit methods that required blasting prior to bench construction and excavation. (Thomas K. Bundtzen)*

wirelike gold strands protrude from quartz. In many other places in Alaska, nuggets and flakes of gold mix with gravel in stream beds. Elsewhere deep in the ground, gold may appear as spidery veins webbing through hardrock and interspersed with silver, copper and lead.

Other minerals take on camouflage too. Garnets protrude from rock ledges near Wrangell in Southeast, yet throughout much of Southcentral and Interior, they tumble with gravel along stream banks. In the Dahl Creek valley near Kobuk, big gray boulders jut from the tundra. Cut open, they reveal shades of jade, green, white and brown, in abstract, variegated designs. Rocky mountainsides near Kennecott take on the colorful hues of at least five types of copper minerals, including chalcopyrite, bornite, malachite and azurite.

Rocks and minerals are products of earth-shaping processes – the volcanoes and earthquakes, glaciation and erosion that made Alaska's mountains, valleys, rivers and oceans.

"Almost every process that takes place, whether from the action of molten rock, heat and pressure at depth, hot springs or steam, running water, weather or biological activity can contribute to formation of an ore deposit," says a U.S. Geological Survey handbook.

"Geologists use the principles of chemistry to try to understand how these processes scavenge elements from ordinary rock, transport them, and concentrate them to form an ore deposit."

Rocks are generally classified in three families — igneous, sedimentary and metamorphic. Certain minerals are associated with each family of rock, and the type of rock determines in what form the mineral might occur.

For instance, chrome deposits at Red

Ore Deposits

Ore deposits are described in terms of their quality or grade, and size or tonnage. The percentage of mineral contained in the ore is its grade; sometimes grade is reported as troy ounces per ton. (Even when not designated, all references to mineral grades are in troy ounces, as opposed to avoirdupois ounces; 12 troy ounces equal a troy pound, while 16 avoirdupois ounces equal a standard, or avoirdupois, pound. One avoirdupois pound is equivalent to 14.581 troy ounces.)

Ore deposits are often referred to as high-, medium- or low-grade. For instance, the Nixon Fork gold-copper deposit grades high at about 1.63 ounces gold per ton. In contrast, an ore grading lower would have less concentrations of metal. A low-grade deposit around Von Frank Mountain, in the Kuskokwim Mountains, averages .013 ounces per ton gold to .035 ounces per ton gold. This gold mineralization, part of a quartz deposit containing copper and iron sulfides and molybdenite, was being explored by ASA Inc. in 1994.

— Adapted from *Alaska's Oil/Gas & Minerals Industry* and *Alaska's Mineral Industry*, 1993 and 1994 reports

A bulldozer moves gold-bearing gravels at this placer mine on American Creek near Eagle. Historically in Alaska, most mining operations have targeted placer deposits. (George Wuerthner)

Mountain near Seldovia probably formed in igneous rocks as magma deep in the Earth cooled. The heavy mineral chromite crystallized and sank to create pods and layers near the bottom of the magma chamber. Today that deposit is being explored by North Pacific Mining Co., a subsidiary of Cook Inlet Region Inc.

In other cases, minerals are suspended in hot, watery solutions, called hydrothermal fluids, which move through cracks and openings, either as the rock is forming or after it has formed. As the fluid cools, the minerals settle out creating veins, threads or even pockets of ore.

The Search Is On

Although gems and minerals have been forming for eons in Alaska, only relatively recently have people come looking for them.

The Russians came mainly for Alaska's furs, but they also found minerals. In 1838, Russian explorers discovered the mercury mineral cinnabar near Kolmokof Fort on the lower Kuskokwim River. In 1848, Peter Doroshin, a mining engineer with the Russian American Co., found gold along the Kenai River near present-day Cooper Landing. But the Russians did little, if any, actual metal mining.

American prospectors followed soon after, and they most definitely were on the trail of gold. Through Southeast came the first few gold diggers, spotting small claims in the 1860s and 1870s at Sitka and at Windham Bay on the mainland. With discovery of the Juneau gold belt in the 1880s, prospecting and mining escalated, a preview of the northern rushes to come.

Fortune hunters started arriving by the shiploads, with news of big strikes farther inland in Alaska and Canada. Hopeful miners scrambled over the coastal mountains through Chilkoot and White passes from Skagway, over the glaciers off Prince William Sound from Valdez, and up the Yukon River from the Bering Sea. Strikes in Alaska's Interior and on the Seward Peninsula put Fairbanks and Nome on the map, towns that are still mining centers today.

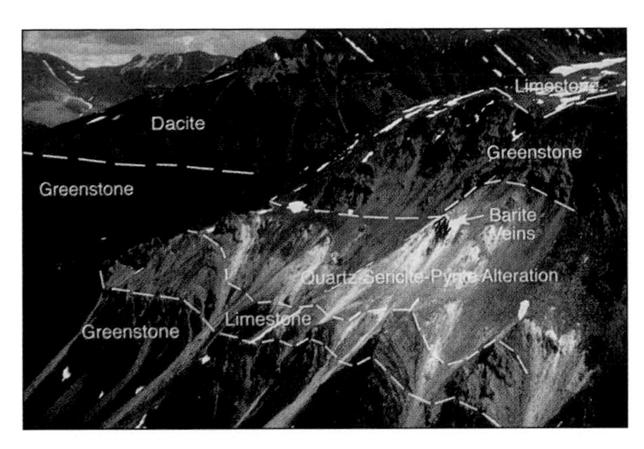

Mountains contain many ore-bearing rock types. This diagram of Toklat prospect in the Talkeetna Mountains shows dacite, an igneous rock of quartz and feldspar with calcium, magnesium or iron components; sedimentary limestone; a type of metamorphosed igneous rocks called greenstone; and a metamorphosed alteration containing quartz, sericite schist, and pyrite iron ores with barite veins. (Thomas K. Bundtzen)

Yet even before this, Alaska's indigenous people used rocks and minerals. They depended on local stones and valued those acquired through trading with other Native groups or, later, foreigners. Pacific Eskimos chipped slate into spear points, made decorative lip labrets of amber and marble, and gazed into polished stone mirrors. The Aleuts used basalt, obsidian, greenstone and chert for their knives, scrapers and projectile points. The Tlingit Indians of Southeast used argillite, a metamorphic shale, to carve bowls and decorate totem poles. They also may have pounded local copper into shields, which they displayed as a measure of wealth; they later acquired thin sheets of European copper, which were intended as ship

Rock Types

Igneous rocks form as molten magma cools and solidifies. If they form deep in the Earth as the magma slowly cools, they are known as plutonic igneous rocks and usually contain a mosaic of small mineral grains. Granite is a type of plutonic igneous rock.

When magma cools quickly, as happens at the surface after a volcanic eruption, it forms volcanic igneous rocks. These may contain large mineral crystals as well as lots of tiny, almost invisible grains. Basalt is a common volcanic igneous rock.

Sedimentary rocks form in layers on the surface of the land or the ocean floor. They may form from erosion of older rocks. When a mountain erodes, for instance, fragments of rocks and soils wash or blow away. They may accumulate elsewhere in a layer and harden, forming sedimentary rock such as sandstone.

Sedimentary rock also forms when minerals precipitate or drop out of a watery solution. Limestone, a type of sedimentary rock, forms when calcium carbonate precipitates from sea water, or from accumulations of sea shells made from calcium carbonate.

Metamorphic rocks form when existing rocks are subjected to heat and pressure deep within the Earth. When limestone undergoes metamorphic processes, essentially getting mashed and cooked in the Earth's kitchen, it turns to marble. Sometimes, metamorphism involves temperatures and pressures so extreme that the existing rock melts, producing magma.

— Adapted from *Alaska's Oil/Gas & Minerals Industry*

LEFT: Early explorers found Alaska Natives from the Copper River area using copper nuggets. (Steve McCutcheon)

ABOVE: Mining continues today around Tofty, a community on Sullivan Creek in the Yukon-Tanana Uplands of the Interior. It originated about 1908 as a mining camp, named after A.F. Tofty who discovered gold in the area and reportedly took out 376 ounces in six weeks. (Arne Bakke)

sheathing. Athabaskans in the Copper River area used native copper nuggets recovered from placer deposits for utensils and bullets. They traded the copper with Indian groups along North America's northwest coast.

Today, Alaska's Natives are partners with mining companies in some of the biggest mining ventures in the state. Red Dog Mine, the largest in Alaska in early 1995, is owned by NANA Corp., the Native regional corporation for the Kotzebue area. The mine is operated by Cominco Alaska, a subsidiary of worldwide mining conglomerate Cominco Ltd.

Most of Alaska's 12 Native regional corporations, including Doyon Ltd., Calista Corp., Sealaska, Koniag, Bering Straits Native Corp., and Cook Inlet Region Inc., are involved with various levels of mineral leasing, exploration, development and mining on their lands.

Throughout Alaska, about 3,150 people identified themselves as miners in 1994. Most mined commercially; only 280 reported themselves as recreational miners.

Then there are people who just like to collect gems and minerals for the fun of it, the rock hounds of Alaska. They spend their weekends panning for gold, walking river beaches looking for agates, jasper and jade, hiking the backcountry in search of quartz crystals, amber or just some interesting mineral specimen to chisel off and take home to admire.

The Land

About 145.5 million acres — slightly less than half of Alaska's total 365 million acres — are federal and state public lands open to mining. Mineral exploration and development also occurs by lease arrangement on some of the 45 million acres in Alaska owned by the Native regional corporations.

Mining, even on private lands in Alaska, is regulated by the state and federal government, which oversee such things as how mining affects water quality and whether the land is returned to adequately good condition when mining ceases. Miners and mining companies are required to obtain mining permits from the government, as well as keep up-to-date mining and reclamation plans on file. In addition, miners must pay certain rents, production royalties, bonds, taxes and other fees as required to the government. Negotiating leases with private landowners comes extra.

While Alaska is nothing if not a giant collection of rocks and minerals, finding an ore deposit worth mining can be tricky.

Consulting records and maps available through the U.S. Geological Survey and the state's Division of Geological and Geophysical Surveys is one starting place. Miners may study geophysical maps or computerized data produced by aerial surveys and ore-modeling programs. The aerial surveys use electromagnetic tools to show surface and underground rock structures indicative of particular types of deposits. Ore-modeling programs show where deposits are likely to occur, based on tools such as ground-water analysis or chemical extractions from surface gravels and soils. Government bulletins may contain geologists' field notes that refer to mineral traces found in streams, rivers, soils or rock outcroppings. And then there is the invaluable and irreplaceable legwork, walking an area and looking for promising signs of mineral occurrences.

Miners also review government records of previous claims in an area. One of the best places to look is archival records maintained by the state's Division of Mining.

Large mining companies may investigate hundreds of sites before finding one potentially large enough to pursue. Their geologists and engineers may spend many thousands of dollars in exploration, to drill and test for ore content. They weigh the economics: the outlook for world mineral production, price and consumption forecasts, mine development and operating costs, and requirements of meeting government permitting and environmental standards.

Alaska's Minerals

No one knows exactly how many different types of minerals exist in Alaska, but most of the minerals and metals used in industrialized society are found here as well as dozens of interest to rock hounds, gem collectors and jewelry makers.

Of Alaska's metals and minerals, approximately 30 are economically significant. In recent years only a handful of them — most notably zinc, gold, silver, as well as sand and gravel — have been valuable enough to pay for mining. Other minerals, such as copper,

Ruins of Franklin, a turn-of-the-century mining camp, overlook South Fork Fortymile River. The post office here closed in 1945. The first big Interior gold strike occurred on the Fortymile in 1886. (George Wuerthner)

platinum and mercury, were extracted by major mines in Alaska's past.

In 1994, Alaska's total mineral production was valued at about $507.5 million, about 13 percent above the previous year. Metals, led by zinc and gold, accounted for more than three-fourths of this value. In addition, mining companies spent more than $30.7 million in exploration and another $45.2 million in development. Added to the value of minerals produced, the total mining industry in Alaska contributed more than $583.4 million.

Here's an overview of the major minerals found in Alaska.

No terrain appeared too daunting for miners as they searched for ore. These prospecting trails hang on the steep and rocky Wrangell Mountains. (Steve McCutcheon)

GOLD

Gold shines as the most sought-after mineral throughout Alaska's history. The early gold miners chased their dreams into the far reaches of the land, sometimes striking it rich but more often eking out only a meager living. From this pursuit emerged a tradition that influenced Alaska's economy, politics and social fabric. Besides that, the exploits of gold rush prospectors added dozens of colorful legends to Alaska folklore.

Gold is found in every region of the state. More than 33.24 million troy ounces of gold have been shoveled, sluiced, panned and dredged in Alaska since 1880 when record-keeping started. By weight, Alaska's production would be equal to about 189 (6-ton) African elephants, the world's largest land mammals. Yet, if those elephants were really made of gold, they'd be the world's heaviest, *tiniest* land mammals. Gold is extremely dense and occupies very little space. All the gold taken from Alaska so far, all those tons of elephants, would fit into a third of a box car on an Alaska Railroad freight train.

Gold is the only mineral continuously produced during Alaska's history.

While Alaska's gold production blasted off with the first Southeast gold mines in the late 1800s, stories of gold also were being carried out of Alaska's Interior by explorers and traders. The first significant Interior strike along the Fortymile River in 1886 was followed by discoveries at Circle in 1893 and along the Seventymile River in 1895.

The 1896 Klondike strike in Canada's Yukon Territory captured the world's attention, drawing thousands of prospectors and entrepreneurs, scam artists and thugs. The time was unlike any other in Alaska's history. The escapades played out on the Chilkoot and White Pass trails out of the gold rush towns of Skagway and Dyea — from Soapy Smith's racketeering to the prospectors who hauled tons of food and gear, including pianos, bowling alleys and flocks of chickens, over the mountains — live on as legends from one of Alaska's craziest periods.

The Klondike rush was followed by more of the same, on a smaller scale. Alaska had 34 back-to-back gold stampedes, including to Koyukuk in 1893, Nome and Council in

1898, Fairbanks in 1902, Innoko in 1906, Iditarod in 1908 and finally to Livengood in 1917.

Athabaskan elder Katie Johns worked at the Nabesna gold camp in the eastern Interior as a young girl and recalls her mother's words in describing the way the prospectors swept into that part of Alaska: "My mother say it's just like a bunch of caribou. All over, everyplace you go."

Alaska's biggest single year of gold production occurred in 1906, when miners extracted more than 1 million ounces. That production was worth $22.04 million in 1906 dollars, a value that wanes compared with gold prices that came later, in the 1980s.

Gold prices peaked in 1980 at $800 an ounce, with a yearly average of about $570 an ounce. The rest of that decade saw prices fluctuate in the $300 to $400 per ounce range, hitting $455 an ounce in 1987. The next year when prices averaged about $425 an ounce, the value of Alaska's gold peaked at $112.84 million for an historically unremarkable production of only 265,500 ounces. Today's average price of gold, at about $385 an ounce, can sustain placer and lode gold mining.

The late-1980s price rise came as a result of world supply and demand, factors constantly influencing the prices for all minerals. In the late 1980s, demand for gold surged, particularly in Japan where gold jewelry sales increased almost 70 percent. Domestic production in the United States increased during that time as well, with Nevada, California, Montana, South Dakota and Alaska the nation's top five gold-producing states in 1990. Alaska has consistently ranked as fifth in the nation's gold production for the last decade.

ABOVE: *Frank and Mike Roberts stand in the portal of their underground drift gold mine on Dome Creek near Fairbanks. Each year, six to eight drift mines have been operated in the Fairbanks, Wiseman, Ophir and Inmachuk areas. About 3,150 people worked as miners in Alaska in 1994. (Thomas K. Bundtzen)*

RIGHT: *Jack Hoogendorn stands at the portal of an underground drift mine south of Deering on the Seward Peninsula. Hoogendorn mined and prospected the area for more than 50 years before passing away in 1992. (Thomas K. Bundtzen)*

In recent years, most of Alaska's gold has come from a few big producers, mostly placers working in the southcentral, interior, northern and western regions of the state. Placer mines, by far, constitute the majority of Alaska mining operations with 185 at latest count. In 1994, placers accounted for all Alaska's gold – 182,200 ounces worth nearly $71 million. Throughout Alaska's mining history, placers have contributed 70 percent of the total gold production.

Here's a look at some of the state's gold-producing regions and largest mining companies. The 10 leading gold producers (see chart, pages 40-41) accounted for 58 percent of the total gold extracted in Alaska in 1994. In addition to the handful discussed here are dozens of individuals and companies doing seasonal or year-round mining, as well as exploration and development.

The state's biggest gold mine for much of its 11-year operation was Cambior Alaska's placer mine at Valdez Creek, off the Denali Highway east of Cantwell. It produced an average of about 40,000 ounces annually. During its peak year of operation in 1992, the mine produced around 86,052 ounces of gold. It employed 150 to 165 workers

ALASKA'S TOP METAL-PRODUCING MINES AND PROSPECTS CURRENTLY UNDERGOING MINE DEVELOPMENT

Company	Mine/Prospect	Location	1994 Employees	Minerals	Year Estab.	Description
Cominco Alaska Inc.	Red Dog	Northern region, near Kotzebue; site owned by NANA Corp.	311	Zinc, lead, silver concentrates	1989	Largest zinc mine in North America; accounted for 6.5 percent of world's mine-produced zinc in 1994 and 58 percent of Alaska's mineral production; 620,590 tons of concentrates shipped 1994.
Fairbanks Gold Mining Inc. (subsidiary of Amax Gold Inc.)	Fort Knox	Interior region, near Fairbanks; on state lands	450 (construction phase 1995)	Hardrock gold	1990	Will be Alaska's biggest gold producer and one of nation's top 10 when mining starts in 1996; construction underway 1995, with 450 employees; will be 250 at full production; mine will process 35,000 tons of low-grade ore daily; estimated 300,000 ounces of refined gold produced annually; reserves are almost pure gold in granite.
Cambior Alaska Inc.	Valdez Creek	Southcentral region, east of Cantwell	151	Placer gold	1984	Largest gold mine in Alaska during 11-year operation, will have produced about 475,000 ounces refined gold by closure in Sept. 1995; biggest year 1992, with 86,052 ounces worth $29 million; Alaska's leading producer in 1994 with 47,622 ounces.
Polar Mining Inc.	Goldstream/ Fairbanks creeks	Interior region, near Fairbanks	54	Placer gold	1981	Alaska's second largest gold producer in 1994; family owned and operated.
Alaska Gold Co.	Nome beaches	Western region	65	Placer gold	1924	Third largest gold producer in 1994 at 17,000 ounces refined gold from two bucketline dredges and one open pit; current dredge operation began in 1974; predecessor company once operated 13 bucketline dredges in four Alaska mining districts, recovered 200,000 to 250,000 ounces of gold annually from 1930s through early 1960s; at least one dredge in continuous production for 71 years except for shutdown during World War II.
Alaska Placer Development Inc.	Livengood/ Tolovana	Interior region	15	Placer gold	1985	Fourth largest gold producer in 1994; surface mine operation.
Silverado Mines Inc.	Nolan Creek	Northern region, near Wiseman	25	Placer gold	1993	Produced gold from underground drift mine in winter; open pit in summer; recovered 41.3-ounce nugget in 1994, unofficially 10th largest in Alaska's history; company intermittently active since 1980.
Little Eldorado Group	Little Eldorado Creek, tributary to Chatanika River	Interior region, north of Fairbanks	7	Placer gold	1994	Modern underground drift mine began in 1994; features large-scale drift mining not seen in Interior for many years.
Sphinx Natural Resources and Malry Technology	Near Ruby	Western region	10	Placer gold	1986	Largest gold mine in Ruby-Poorman District for about 10 years; winding down operations in 1995.
Nyac Mining	Tuluksak River drainage	Southwest region, lower Kuskokwim	15	Placer gold	1973	Started as Tuluksak Dredging Co., working on Bear Creek 1973 to 1987; new company formed in 1989 to operate at present location using open cut methods; had employed up to 30; now winding down.
Taiga Mining	Hogatza River	Northern region	7	Placer gold	1990	Former employees of Tuluksak Dredging Co. acquired historic dredge from Alaska Gold Co. that previously worked Hog River area from 1957 to 1975 and was Alaska Gold's main gold source during those years.

Company	Mine/Prospect	Location	1994 Employees	Minerals	Year Estab.	Description
Thurmond Oil & Mining	Eureka-Manley	Interior region	15	Placer gold	Mid-1980s	Dependable Alaska producer for last decade, working one to two open cuts.
Green Mining & Exploration	Ruby	Western region	10	Placer gold	1970	Long-time Alaska producer, frequently ranking in top 10 gold producers.
GHD Resources	Tofty	Interior region	15	Placer gold, some tin	1983	Decade-long operation that's successfully worked mines in Circle, Candle and Tofty districts; shutdown in 1994; owners bought into Little Eldorado Group.
Paul & Co.	Circle	Interior region	10	Placer gold	Late-1970s	Has been one of the Circle District's largest gold mines; currently scaling down.
Rosander Mining Co.	Colorado Creek	Western region, north of McGrath	5-10	Placer gold	1975	Largest producer for many years in Innoko District; has operated off and on for 50 years; started by Toivo Rosander, a Finnish immigrant who came to Alaska in the 1930s; son Ron and grandson T.J. in charge of operations; complete skeleton of woolly mammoth recovered and donated to University of Alaska.
Kennecott Greens Creek Mining Co.	Greens Creek	Southeast region, Admiralty Island	50 (during shutdown)	Silver, gold lead, zinc	1989	Nation's largest silver producer until low ore prices idled mine in April 1993; expected to resume operation 1997 with 250 employees and expanded ore base; seeking agreement with U.S. Forest Service to extend underground operations into Admiralty Island National Monument in exchange for rent and royalty payments.
Nevada Goldfields Inc. (subsidiary of Consolidated Nevada Goldfields)	Nixon Fork (near McGrath)	Western region	35 (estimated)	Hardrock gold & copper	1986	High-grade complex expected to produce by late 1995; small underground gold operation will mine 150 tons of ore daily, about 50,000 ounces refined gold annually with copper by-product; exploration started 1983 by Battle Mountain Gold Co. of Nevada; leased to Central Alaska Gold Co. of Fairbanks in 1989; rights to mine property sold to Nevada Goldfields in 1993.
Echo Bay Exploration	Alaska-Juneau	Southeast region	50-80	Hardrock gold	1985	Hopes to reopen historic underground mine; still working on final permits; project delayed over issue of tailings disposal; could employ 450; production estimated at 300,000 ounces of gold annually, processing 22,500 tons of low-grade ore daily.
Coeur-Alaska	Kensington & Jualin	Southeast region	25	Hardrock gold	1985	In predevelopment; Coeur Alaska purchased Echo Bay's partial ownership in Kensington in 1995; annual production projected at 200,000 ounces from Kensington.
LaTeko Resources Inc.	Ryan Lode & True North	Both Interior region, Ester Dome & Pedro Dome, near Fairbanks	25	Hardrock gold	1985	Both in exploration and development in 1995; small-scale heap leach produced about 31,000 ounces of gold-silver bullion from 1986 to 1990 at Ryan Lode; experiment discontinued, now in remediation; Ryan Lode shear zone contains 822,000 ounces of mine-recoverable gold; True North on historic Old Hindenburg gold property; Amax Gold explored in early 1990s; optioned to LaTeko in 1994, in 1995, Newmont Mining working on agreement to develop True North deposit.
USMX Inc.	Illinois Creek	Western region on state lands	4	Hardrock gold, Polymetallic	1994	In predevelopment; would produce 50,000 ounces gold and silver by-product annually.

Independence Gold Mine near Hatcher Pass in the Talkeetna Mountains produced 141,000 ounces of gold during its peak years of operation, 1936-1942. It reopened briefly after World War II, but shut down permanently in 1950. Today it is part of Independence Mine State Historical Park, open in summer for walking tours. (Harry M. Walker)

year-round, except when winter temperatures plunged below minus 40 degrees and heavy equipment refused to operate.

The mine opened in 1984 to exploit paystreaks found on several ancestral drainage channels to Valdez Creek. It closed temporarily in October 1989, but reopened in 1990 with changes to operate more cost effectively. The mine was scheduled to shut down in September 1995, when the available ore reserves were depleted.

Much of Alaska's gold historically came from the gold-rich Interior, the leading gold-producing region in the state most years. Today, Interior miners still dig out the precious metal in quantity.

The Interior boasts the state's newest and soon-to-be-largest gold mine at Fort Knox, 15 miles north of Fairbanks. The Fort Knox Mine, operated by a subsidiary of Amax Gold Inc., started mine construction in summer 1995 with production planned by 1996. Fort Knox's gold reserves are estimated at 158 million tons for an annual production of about 300,000 ounces of refined gold at full operation. (See article on page 64.)

Several other of the state's top-producing gold mines are located in the Interior as well. Polar Mining Inc., a family owned placer operation, ranks as one of the state's top gold producers, mining large open cuts on Goldstream and Fish creeks. Other steady Interior producers have been Cooks Mining,

BELOW: *This 41.3 ounce gold nugget, unofficially the 10th largest in Alaska's history, was recovered in 1994 by Silverado Mines Inc., from their underground placer operation near Wiseman. Although the busiest period of placer mining occurred early this century, three of the state's 10 biggest nuggets have been recently discovered, according to Tom Bundtzen. These rankings are "unofficial," he says, since other, larger finds may go unreported. In 1984, a softball-sized nugget weighing 91 ounces, thought to be the fourth largest Alaska nugget, was found on Glacier Creek in the Kantishna area by local miner Mic Martinek. A 121-ounce nugget, the state's third largest, was recovered in 1986 on Ganes Creek west of McGrath by Magnuson Mining Co. The biggest gold nugget at 172 ounces came from the flanks of Anvil Mountain near Nome. The second largest, a 139-ounce nugget, was taken from the Hammond River near Wiseman in 1914. (Silverado Mines Inc., courtesy of Thomas K. Bundtzen)*

Gold

About 84 percent of the world's gold is used in jewelry production. Rarely is gold jewelry made of the pure metal, which would be defined as 24-karat gold; more often, gold is combined, or alloyed, with other metals. This gives gold the hardness needed for jewelry manufacture and allows for a variety of colors. Copper gives gold a pinkish tinge, while alloys of gold, nickel, palladium and zinc produce white gold. The yellow gold routinely used in jewelry is actually an alloy of gold, silver and copper.

Gold's properties make it an important metal today in telecommunications, electronics, space, medicine, for coinage and jewelry. It is extremely malleable; one ounce can be stretched thin enough to wrap around the globe. It can be hammered into tissue-thin sheets. Only silver and copper conduct electricity better.

The electronics industry uses more than 1 million ounces of gold annually to coat telephone jacks and dialing contacts, and more than 95 percent of the electrical contacts and integrated circuits in computers are gold-coated. Gold is used in the aerospace industry to reflect heat and protect from x-ray and solar radiation. Medical researchers are investigating the use of gold in treating AIDS, gout and other diseases; already gold compounds are effective in treating rheumatoid arthritis.

—From "Nonferrous Metal Outlook 1994, Gold," by Gilles Couturier, Natural Resources Canada, and *The Concise Columbia Encyclopedia.* (Photo by Steven Seiller)

Alaska Gold Co. and its predecessors have operated since 1924 in the Nome and Fairbanks areas. At left, dredge No. 5 works within sight of the Bering Sea, helping make the company Alaska's third largest operator in 1994. Above, a water piping system thaws permanently frozen ground so the dredge can operate. (Both, Steve McCutcheon)

GHD Resources, Alaska Placer Development, the Little Eldorado Group and Thurman Oil and Mining Inc.

The Interior continues to have dozens of small mining operations, and a considerable number of larger companies, in serious exploration and development phases. These include LaTeko Resources Corp., exploring and considering development of several promising deposits north of Fairbanks. The Ryan Lode and True North deposits are their two best prospects.

Another big gold producing region is western Alaska. Here sits the historic gold rush town of Nome, today a community of about 3,500 people that serves as an business and commercial center for the region.

Miners first struck gold here in 1898 on Anvil Creek, about a dozen miles inland from Norton Sound. Within a year, 10,000 people lived in the mining camp, then called Anvil City. When word hit Outside, thousands more arrived on ships from Seattle and San Francisco. The city of Nome ballooned to 30,000. Mining fever spread along the Norton Sound beach, where in two months some 2,000 miners removed about 50,000 ounces of gold.

The *Nome Nugget* described the Nome beach as "the world's largest junkyard" with "a lot of jackass machinery." Miners rigged hundreds of digging machines, some at great costs, that were abandoned when sea swells and storms rendered them useless.

Prospectors also staked claims on creeks throughout the area. At least four major placer deposits were discovered inland on former coastlines.

The Nome District contributed 350,000 ounces of gold to 1906's record year of production in Alaska. By 1910, electric dredges were commonplace. Except for the hiatus during World War II and a 12-year break during the 1960s and early 1970s when gold prices dived, onshore gold dredging has been part of the Nome mining scene.

Most recently, the Alaska Gold Co., the largest producer in western Alaska, operated two big onshore dredges. The company plans to convert the dredging operation to open pit mining starting in 1995.

Summertimes today still find visitors

optimistically working Nome's beaches with shovel and sluice box, and maybe small pumps. Some of them find a bit of gold, and after a storm it's possible to see the thin, horizontal lenses of gold- and garnet-containing sands and heavy gravels.

From 1986 through 1990, the world's largest offshore bucket-line dredge, the BIMA, operated here.

New hardrock gold mines under development in the western region may further boost Alaska's gold production when they go on line. Consolidated Nevada Goldfields appears on the fast track to develop an underground mine and mill at the Nixon Fork high-grade gold and copper deposit northeast of McGrath. The company hopes to achieve a production rate of 60,000 ounces of gold per year, once the mine opens. Production is set to start in October 1995.

At Illinois Creek, USMX Inc. is trying to develop a gold-polymetallic deposit and estimates future production at about 50,000 ounces of gold a year.

The oldest gold-producing region, Southeast, was where gold mining began in Alaska. It started small, as prospectors trickled into Southeast out of Canada and up the coast from California, disappointed they hadn't struck rich pay during the 1848 California gold rush or the 1873 stampede to Cassiar, B.C. They found small deposits near Sitka and at Sumdum and Windham bays on the mainland.

In 1879, naturalist John Muir noted promising mineralization along Gastineau Channel. A mining engineer in Sitka, George Pilz, put out the word that he'd pay handsomely in blankets to any Native who brought him paying ore. A Tlingit named Kowee did just that, and Pilz grubstaked a

Marine Mining

Offshore dredging to mine the riches thought to cover the sea floor was first attempted in the Nome district in 1900 at the mouth of the Snake River; a storm destroyed the dredge only days after it started operating. Other efforts continued. In 1905, a miner hired divers in winter to descend under the sea ice and bring sediments to the surface.

Later submarine mining efforts included using a clam-shell scoop working from offshore ice in winter to lift sediments through a hole in the ice. Other imaginative schemes proposed in the early 1960s included a submarine dredge that would crawl along the sea floor.

But for the most part, the district's possible offshore riches didn't attract serious study until Alaska achieved statehood and acquired jurisdiction three miles from shore. The first offshore prospecting permit was issued to Shell Oil Co. in 1962; others soon followed. In 1967, the federal Bureau of Mines started an offshore drilling program. Together, the explorations confirmed existence of a major offshore placer. Weather and rough seas, the price of gold, and the lack of offshore dredging technology hindered development. Local entrepreneurs experimented with ideas such as modifying a road grader to mine the shallow seabed deposits, but it floated away on its big balloon tires. Development of offshore mining technology taking place internationally eventually came to Nome.

The BIMA (shown here), the largest bucketline dredge in the world, was designed to work offshore tin deposits in Indonesia. When tin prices fell, Western Gold Exploration and Mining Co., which held seabed leases near Nome, bought the BIMA and towed her to Nome in 1986. The BIMA produced about 120,000 ounces of offshore gold before it left Alaska in 1990. In addition to extracting gold, the Bima's operation expanded knowledge about offshore sediments, gold grades and other important underwater features.

— Adapted from Richard Garnett, *The Alaska Miner*, July 1992. (Photo by Thomas K. Bundtzen)

couple of prospectors to check out Kowee's site in the Gastineau area. They returned with a half-ton of gold ore in their canoe, setting off a small rush out of Sitka in winter 1880 to the beach of what would become Juneau. By the next spring, the frontier gold camp was getting monthly steamship service and miners were producing steady pay from placer and lode claims in the surrounding streams and valleys.

The lode claims belonging to John Treadwell and his San Francisco partners, on Douglas Island across the channel from Juneau, were particularly rich. The Treadwell complex, composed of the Treadwell, 700 Foot, Mexican and Ready Bullion mines, grew into an empire that dominated Gastineau Channel for 35 years, writes author Sherry Simpson in *ALASKA GEOGRAPHIC®*'s *Juneau* issue (1990). The federal government shut down the gold mines in 1942, because of World War II.

The Treadwell complex was the first big commercial gold operation in Alaska and one of the largest underground operations in the world. The ore was not high grade, but enormous quantities were processed. From 1887 until 1922, Treadwell produced more than 3 million ounces of gold from 28.8 million tons of ore, employing some 2,000 miners in three shifts almost every day of the year. Then in 1917, a disastrous cave-in flooded three of the mines. Incredibly, none of the miners died; they were evacuated safely as sea water filled the mine shafts and gushed out the entrance portal in a 200-foot spout. The Ready Bullion Mine alone was untouched and continued producing gold until 1922, when low prices finally closed it.

Mining continued in Juneau, however, at the Alaska-Juneau (A-J) on Mount Roberts behind town. The A-J was the most successful venture of a company that had worked area lodes since 1897. In 1910, the company started tunneling the Gold Creek Adit into

LEFT: *This 1942 aerial of Gastineau Channel shows piles of waste rock and tailings from the Treadwell mines (white, lower bank) and from the Alaska-Juneau Mine (upper bank) used to fill in tidelands and expand the city of Juneau. (Steve McCutcheon)*

BELOW: *A prosperous Juneau dominates this view from the Alaska-Juneau Gold Mine tram in 1942. (Steve McCutcheon)*

ABOVE: *Hydrologists from the state Department of Natural Resources monitor water quality of streams throughout the state affected by mine activities. Mine operators are required by the state to treat mine wastes and effluents to meet acceptable water quality standards. (Thomas K. Bundtzen)*

RIGHT: *Strikes in Southeast and the Interior first lured adventurous gold seekers to Alaska in the 1880s. Promoters cashed in on the escalating gold fever, often using unscrupulous tactics such as spreading unfounded stories of new finds and advertising unexplored routes. Such as this landed some 4,000 prospectors at the foot of Valdez Glacier in winter 1898, duped by promises of a swift, easy crossing to rumored gold-rich Copper River country. Many of them returned to Valdez destitute and sick in a mass exodus the next fall. Gold was later discovered near the Valdez camp; the largest producer was the Cliff Mine staked in 1906 on Shoup Glacier, 10 miles from town. These unidentified miners worked in the area. (P.S.Hunt, Anchorage Museum, Photo No. B62.1A.52)*

the mountain. By 1917, its new mill to process the ore was running. In 1924 the operation finally started showing a profit, processing tons of low-grade ore, averaging about 13,000 tons a day at peak production. Waste rock dumped into the channel created a harbor for Juneau and was used to make roads and fill in tidelands for the city's expansion; finely ground tailings discharged into Gastineau Channel formed a sandy beach.

The A-J operated on a reduced basis during early years of the war, because of its importance to Juneau's economy, but then closed permanently in 1944. It historically was the largest lode gold producer in Alaska.

Today, Juneau's mining legacy is peddled as a visitor attraction, with museum displays and tours to the old mine sites. But mining isn't dead here, if developers succeed with plans to reopen the historic A-J, Kensington and Jualin gold properties.

Echo Bay Alaska Inc. has been preparing for more than a decade to reopen the A-J. The company predicts milling 22,500 tons of ore a day, producing 375,000 ounces of gold annually, if the mine is finally permitted to open. Echo Bay continues exploration to expand the known reserves of 100 million tons. In conjunction with the mine, the company has proposed building a $6 million tourist facility, with historic displays and archives and closed-circuit video monitors

ABOVE: *The tiny village of Nabesna, at the base of White Mountain in the Wrangells, began in the early 1900s as a gold camp for the Nabesna Mining Co. (George Wuerthner)*

TOP RIGHT: *Gold washed out through sluicing is being collected during sluicebox cleanup, generally two to five times a year, at Rosander Mining Co.'s placer operation on Colorado Creek, north of McGrath. (Thomas K. Bundtzen)*

RIGHT: *Polar Mining Inc., Alaska's second largest gold mine in 1994, operates year-round, even when temperatures drop to minus 40 degrees at its lower Goldstream Creek placer mine in the Interior. Stripping and ore stockpiling occur in winter; sluicing in summer. (Thomas K. Bundtzen)*

that would allow visitors to see miners at work inside the mountain, extracting ore and pouring the purified metal into bricks.

But the project has been delayed, because of opposition from some Juneau residents and environmental groups. Their objections have focused on the company's plans for disposing of mine tailings. Echo Bay proposed building a dam to flood nearby Sheep Creek valley to use as a mine tailings disposal pond. Objectors say the valley is popular for recreation such as hiking and shouldn't be lost to industrial development.

In reviewing the Sheep Creek proposal, the federal Environmental Protection Agency found problems with possible surface water draining into and out of the pond, because of the valley's steep terrain. The agency is looking at changing its policies to allow Echo Bay to dispose of the tailings in the deep water of Taku Inlet. Meanwhile, company officials are trying to work with environmental groups to reach some type of understanding.

At the Kensington Mine about 45 miles north of Juneau, Coeur Alaska Inc. gained full operating control when a buy-out of Echo Bay, a former partner in the venture, was completed in summer 1995. The Kensington gold grades higher than that at the A-J, but the reserves are considerably smaller, only about 13.5 million tons. Coeur spokesmen at the company's headquarters in Idaho have said that construction at Kensington may start as early as 1996, once the project receives final permitting.

Coeur also is looking at the Jualin property, on the southeast side of the mountain containing the Kensington. Its gold grade is even higher, but its reserves are only about 900,000 tons.

Another big gold producer in Southeast, the Greens Creek Mine, wasn't built as a gold mine at all, but instead as a poly-metallic metals mine targeting silver. The mine opened in 1989, producing significant amounts of gold as a by-product until it closed in 1993 when metal prices dropped.

SILVER

Silver has been produced in Alaska almost every year since 1880 for an historic total of nearly 61.2 million ounces. Virtually all of Alaska's silver ores occur as complex minerals of silver, lead, zinc and antimony in fractures cutting a variety of rock types.

Most of Alaska's silver was produced as a by-product of copper and gold mining until 1989, when Greens Creek started production.

In its first year of production, Greens Creek extracted more than 5.2 million ounces of silver, making it the biggest silver mine in the United States. In subsequent years, its silver production increased to more than 7 million ounces annually. Through April 1993 when it closed, the mine produced 29.23 million ounces of silver, or 48 percent of all the silver ever mined in Alaska. For four of its five productive years, it ranked as the nation's largest silver mine. The mine also produced gold and concentrates yielding lead and zinc.

Before Greens Creek, most of Alaska's silver came out of the Kennecott copper mines in the Wrangell Mountains earlier this century. (See Copper, page 52.)

Sources of primary silver in the past came from ore deposits in the Omilak area of the Seward Peninsula north of Golovin; the Kantishna, Fairbanks and Ruby districts of the Interior; the Kaiyuh District in the western region; and the Riverside Mine in Southeast.

The largest producer of primary silver ores was the Kantishna District on the north flank of the Alaska Range during a time of high silver prices following World War I.

Because most of the ore deposits containing silver in the state require expensive underground mining methods, production has been limited to times of high silver demand.

Silver

One of the first metal minerals ever used by humans, silver is an excellent conductor of heat and electricity. Pure silver is nearly white, lustrous, soft and very malleable. Silver is used in coins, mirrors, utensils, antiseptics, jewelry, plated and sterling tableware, dentistry, electrical contacts and photographic emulsions. Sterling silver is an alloy, 92.5 percent silver and 7.5 percent copper.

Ore minerals containing silver include the silver sulfosalts – argentite, cerargyrite, pyrargyrite, stephanite and proustite.

World silver demand is affected by the activities of silver-hording countries, chiefly India which is effectively on the silver standard. For example, when India sold some of its silver stocks in 1992 and 1993, the price collapsed, causing in part the closure of Greens Creek Mine. India's silver reserves are measured in the billions of ounces.

– Adapted from
The Concise Columbia Encyclopedia

Where did Nome's beach gold originate? Early observers suggested volcanoes in the ocean belched gold, or that gold came from meteor showers, or that the Bering Sea floor was covered in gold that the waves washed ashore, writes Richard Garnett in a 1992 issue of The Alaska Miner. In truth, the beach was a natural collection site for gold eroded from the uplands; the metal was concentrated on the beach by surf action. Gold was emplaced in the uplands thousands of years ago by the southerly passage of a major glacier that dumped gold-bearing debris. Today, recreational gold miners, like Andy Hehnlin from Girdwood, Alaska work a 40-mile stretch of the Norton Sound coast between Hastings Creek and Cripple River. They operate on public beach between low and high tide marks; state placer mining permits are required to dredge offshore with floating equipment carried on the yellow pontoons. A hard day's work of beach placer mining with hand tools can yield as much as two ounces, worth about $750 at 1994 prices. (Thomas K. Bundtzen)

Zinc

The automotive and construction industries depend on zinc. About 48 percent of the world's zinc in 1993 was used to galvanize steel, the most cost-effective method of protecting steel against corrosion. Galvanized steel is used extensively in cars and for roofing and siding. Zinc-aluminum coatings provide corrosion protection for large steel structures such as bridges and hydroelectric transmission towers. The high cost of lumber is making zinc-coated structural steel cost-competitive for home construction, which rose from 500 steel-framed homes in the United States in 1992 to 75,000 by 1994.

The manufacture of brass and bronze is the second most important use of zinc. These alloys are used in plumbing fittings, heating and air conditioning components and other products. The third most important use of zinc is for die-casting products such as builders' hardware and automobile fittings.

— Adapted from "Nonferrous Metal Outlook 1994, Zinc" by Philip Wright, Natural Resources Canada

ZINC

The bluish-white element zinc occurs widely in Alaska in a zinc-sulfide ore called sphalerite.

The best known zinc-bearing ores in Alaska are the copper-zinc, massive sulfide deposits along the southern flank of the central Brooks Range and the zinc-lead deposits of the De Long Mountains, notably the Red Dog and Lik deposits. The first commercial production of zinc in Alaska occurred in 1947 from a tiny mine at Mahoney Creek near Ketchikan. A few hundred tons of zinc were shipped from this mine during its three years of operation.

Largely due to development of the Red Dog Mine, Alaska has become the nation's largest producer of zinc. Since it opened in

The Greens Creek Mine on Admiralty Island closed in April 1993 after a period of sliding metal prices, putting about 200 miners out of work and stunning the economy of Juneau where most of the miners lived. Greens Creek is a joint-venture partnership of Kennecott Corp. of Salt Lake City and Hecla Mining Co. of Idaho.

Underground exploration by operator Kennecott Greens Creek Mining Co. continued during the shutdown. The company plans to reopen the mine by January 1997, after spending millions to develop a southwest extension of the orebody on its property. The mine, located 25 miles west of Juneau, was the city's biggest private employer with annual payrolls of $10 million to $13 million. Most of the workers commuted daily to the island by ferry. A skeleton crew of about 25 people kept on at the mine had grown to about 70 in 1995, and the company planned to bring the work force up to about 250 to reopen.

The company hopes to expand underground drilling off its property into adjacent Admiralty Island National Monument through a land swap with the U.S. Forest Service. In this agreement, subject to Congressional approval, Greens Creek will receive 7,500 acres of subsurface mineral rights, in return paying the Forest Service $1 million plus mineral royalties. The Forest Service will use the payments to buy private inholdings on Admiralty and elsewhere in Southeast. The mine lands will revert to the Forest Service when operations end.

The Greens Creek sulfide deposit, part of a mineralized belt cutting diagonally across the island northwest to southeast, was first explored in 1973. Preliminary engineering started in 1981 and by 1989, the mine and mill opened at a cost of $114 million, significantly more than the $80 million initial estimate.

During the mine's first five years of production, it produced a yearly average of 7.6 million ounces of silver, 37,000 ounces of gold, and 58,700 tons of lead-zinc concentrates. (George Wuerthner)

1989, Red Dog has become North America's largest zinc producer, accounting for up to 64 percent of the nation's output. It is also a major player in the world zinc market, supplying 6.5 percent of the mine-produced zinc in 1994.

The mine also contributed 58 percent of the total value of Alaska's 1994 mineral production. It employed one of the largest work forces in the state, with 311 full-time jobs producing more than 550,000 tons of zinc, lead and imperial smelter feed (a combination of zinc, lead and silver) concentrates a year.

Red Dog still has substantial reserves — more than 76 million tons averaging 17 percent zinc, 5 percent lead and 2 to 5.5 ounces per ton silver — to operate for the next 40 to 50 years. The mine ships its ore concentrates to overseas and Canadian refiners from a port at Kivalina.

Sometime in 1993, the one-millionth ton of Alaska zinc, mainly from the Red Dog Mine, left shore for overseas markets. About 10 percent of Alaska zinc output has been as a by-product from the Greens Creek Mine in Southeast.

In the early 1990s, Cominco made changes to improve the Red Dog mill output and metallurgical recovery to counter falling metal prices. Because of this, its zinc concentrate shipments increased more than 40 percent in 1994, to an output for the year of 659,000 tons, or 55 percent of U.S. mine production of zinc. At the same time, the mine froze wages, until zinc prices increase. Meanwhile, low zinc prices halted 22 other zinc operations internationally, decreasing the overall supply.

However, the use of galvanized sheet steel in automobile manufacturing is increasing, from 180 pounds per car in 1981, to 590 pounds per car in 1991, and 1,000 pounds per car by 1995. This should mean increased demand and increasing prices, with Red Dog poised to take advantage of the market.

Near Red Dog is the Lik zinc deposit. Echo Bay Alaska, a joint owner of this deposit, is doing exploration work and looking at the possibility of sharing Red Dog's access road and port facilities if Lik is developed. Lik's 21.7 million tons of reserves grade 9 percent zinc, 3.1 percent lead and about 1.5 ounces per ton silver.

LEFT: *Geologists from Kennecott Corp. and the state Division of Geological and Geophysical Surveys examine an exploration trench in the Nome District. Mining companies may investigate hundreds of sites before finding a prospect valuable enough to develop. (Thomas K. Bundtzen)*

ABOVE: *The mining industry uses helicopters to fly supplies into remote Alaska locations during exploration and development. Helicopters also figured prominently into a statewide aerial geophysical mapping program, the first in 20 years that was launched in 1993 by the state Division of Geological and Geophysical Surveys. The surveys precipitated a flurry of activity as miners rushed to stake new claims. Mapped areas included 494 square miles around Nome, 338 square miles near Circle, 183 square miles around Nyac, and 75 square miles near Valdez Creek. In February 1995, the state released surveys of 742 square miles in the Fairbanks and Richardson districts; more than 80 square miles were staked in four months. Special equipment carried by helicopter measured electrical, magnetic and conductive properties of minerals in the ground below. The collected data is used to map the underground topography and can show various views of rocks and rock formations beneath the surface when enhanced by computer. The resulting charts may reveal structures that look similar to known ore deposits. It's much too early to know how many, if any, new mines will come on line from this. (Thomas K. Bundtzen)*

Copper

Copper is the best conductor of electric power, signals and heat of any of the industrial metals. Its superb conductivity, strength and resistance to corrosion make it valuable to electrical, transportation and construction industries. The chief commercial uses of copper and copper alloys are electrical wire, water tubings, castings, roofing, utensils, coins, metalwork, and in refrigeration coils and heat exchangers. Copper compounds are used as insecticides, fungicides and paint pigments. Promising new markets for copper include natural gas systems, solar power generation and storage of spent nuclear fuels. Copper ores include chalcopyrite, and bornite, shown here. World copper demand has been relatively high in the last decade, fueled by industrial growth in Asia, particularly in China.

—Adapted from "Nonferrous Metals Outlook 1994, Copper," by Geoff Bokovay, Natural Resources Canada, and *The Concise Columbia Encyclopedia.* (Photo bySteve McCutcheon)

Alaska's famous Kennecott copper mine is a popular tourist attraction in Wrangell-St. Elias National Park and Preserve. One story about the discovery of Kennecott's deposits says prospectors mistook green-colored outcroppings on the mountainside as grass for their horses and found copper ore instead; another says prospectors washing dishes in a stream found a copper nugget and followed the water to its source at the orebody. (Penny Rennick)

COPPER

For a brief time early this century, copper reigned in Alaska mining. Copper commanded the highest price of any mineral between 1916 and 1920, and the territory became one of the nation's top 10 copper producers.

Alaska's affair with copper started back in 1900, with the first commercial copper mines on Prince of Wales Island. Most of the production came from three mines working copper-iron-zinc deposits. In 1905, two copper smelters were built, one at Hadley on the Kasaan Peninsula and the other at Coppermount on Hetta Inlet. They operated intermittently until 1908, when declining ore quality, falling prices and problems getting coal for the smelters forced them to close.

Then copper mining shifted to Prince William Sound. The Beatson Mine on Latouche Island had started shipping copper in 1904 and dramatically increased production in 1914, when Kennecott Copper built a concentrator capable of handling 1,500 tons of ore a day. By 1918, the Beatson was the territory's largest mine, extracting 435,826 tons of ore a day.

Another major Prince William Sound copper mine was Ellamar, 20 miles southwest of Valdez. It operated from 1902 until 1919, when mining its low-grade ore became unprofitable.

Company towns built up around both operations, thriving and dying with the fortunes of the mines.

The most famous Alaska copper mine,

and one of the richest in the world at the time, was the Kennecott complex in the Wrangell Mountains, which operated as two distinct mines feeding a central concentrator from 1911 to 1938. Two of its orebodies — the Bonanza and Jumbo — held extraordinarily high copper grades, 50 percent and 70 percent respectively. Its owners, the Guggenheims, were one of America's wealthiest, most powerful families. They controlled an industrial dynasty that included American Smelting and Refining Co., a dynasty they expanded into Alaska with shrewd acquisition of numerous properties, including Kennecott.

To do this, one of the Guggenheim sons, Daniel, created the Alaska Syndicate with financing by international banker J.P. Morgan and Co. The syndicate held salmon canneries, two steamship companies, a trading company, coal lands and mines, including Kennecott Copper Co., and opposed any extension of home rule that could threaten its power block. This angered Alaskans, and newspaper editorials vilified the "Morganheims" for monopolistic greed.

Anxious to export Kennecott's copper to their outside smelters, the syndicate pushed through a 200-mile-long railroad to connect Kennecott with the ocean at Cordova, muscling out other railroad developers in the process. The daring Copper River & Northwestern Railway, a $28 million project engineered by ace railroad builder Michael Heney of White Pass Railroad fame, bridged surging glaciers and hung along steep-sided canyons. It's goal: To reach what novelist Rex Beach called the "richest copper district in the world," in his 1912 bestseller *The Iron Trail.*

At Kennecott, aerial tramways hauled ore to the 13-story stamp mill, built into a mountainside. About 300 people lived in the company's town, while the miners lived up the mountain outside the mine entrances. Today the old town site and mill are main tourist attractions in Wrangell-St. Elias National Park and Preserve.

The mines extracted a staggering amount of high-grade ore – 4.6 million tons that averaged more than 13 percent copper and about 2 ounces per ton of silver. From this, it produced 591,535 tons of copper and 9 million ounces of silver valued at $287 million. The company's profits of about $100 million enabled acquisition of mines in New Mexico, Nevada and Arizona.

Today, Kennecott is one of the world's largest metal mining companies, still with interests in Alaska, most notably at Greens Creek. In the 1960s, Kennecott explored and partially developed a copper mine near Bornite in northwestern Alaska, and in the 1970s, the company explored a major copper deposit at Arctic, in the southern Brooks Range.

In 1977, the U.S. Geological Survey reported that a 25-mile-long belt in the Wrangells may contain as many as seven undiscovered orebodies as large as Kennecott's Jumbo. Nothing of that caliber has yet to be found, however.

More recently, Cominco Alaska announced the discovery of a major low-grade copper deposit known as Pebble Copper, about 20 miles north of Iliamna village.

Despite promising discoveries and a healthy copper market, no copper is currently being mined in Alaska.

American Copper and Nickel Inc. drilled a copper prospect in this remote location on the Alaska Peninsula, south of Lake Iliamna, in the early 1990s, then decided it wasn't economical to develop. (Thomas K. Bundtzen)

Other Polymetallic Prospects

Companies continue exploring polymetallic deposits in Alaska.

Westmin Resources Ltd. of Vancouver recently did exploratory drilling at the Johnson River polymetal deposit west of Cook Inlet at the foot of Iliamna Volcano. This deposit, owned by Cook Inlet Region Inc., shows gold, silver, zinc, lead and copper reserves.

Northern Alaska saw a flurry of exploration activity in 1994 in the Ambler copper-zinc belt. Four companies, NANA Corp., Cominco Alaska, Kennecott, and Teck Corp., were active with Teck staking 160 new claims.

In western Alaska, Cominco staked 120 new claims in the Aurora Creek polymetallic zone.

On the west side of Cook Inlet at the Pebble Copper deposit, Cominco has defined a 450-million-ton deposit of about two-thirds copper and one-third gold. The project is apparently on hold until copper prices stabilize.

American Copper & Nickel Co. has been exploring for polymetallic deposits on the north side of the Alaska Range near Tok in the Interior, as well as in Southeast at its Hetta Inlet property on Prince of Wales Island.

Iron

A silver-gray metal, iron is the fourth most common element in the Earth's crust, and possibly the main component of the Earth's core. While a good conductor of heat and electricity, its magnetic properties — both in attracting magnets and in being easily magnetized — are remarkable.

Iron ore minerals include iron oxides magnetite and hematite, as well as limonite, taconite and siderite. A distinctive and common iron mineral is the iron sulfide, pyrite, found in almost every geological environment, but rarely exploited except to produce sulfuric acid.

Iron ores are refined in a blast furnace to make pig iron, which can be remelted and poured into molds to make cast iron, commercially purified to make wrought iron, or alloyed with carbon and other elements to make steel. Iron compounds are used as paint pigments, in dyeing and ink manufacture. Iron is a component of red blood cells in vertebrates; a lack of nutritional iron can result in anemia.

— Adapted from "Nonferrous Metals Outlook 1994, Iron," by Bruce Boyd, Natural Resources Canada, and *The Concise Columbia Encyclopedia*

IRON

Alaska has many occurrences of iron ores, but none of sufficient grade, tonnage and accessibility to be mined. Production from Alaska's iron deposits probably depends on the steel markets of Pacific Rim countries, and transportation costs from Alaska will figure into developing iron ore deposits.

The United States is a major iron ore producer, chiefly from ores near Lake Superior. Titanium has been produced from beach sand deposits in the eastern United States.

Iron deposits in Southeast Alaska, at Klukwan and at Snettisham south of Juneau, have drawn periodic attention for the last 30 years. Japan has been mentioned as a market, but there are no plans for development. The high titanium content of the magnetite, which is disseminated through igneous rock, poses technical problems during smelting.

For a time in the early 1990s, Midrex Corp., a subsidiary of Kobe Steel Corp. of Japan, proposed building an iron reduction plant in Alaska. The company first considered locating in the emerging port complex at Point MacKenzie on upper Cook Inlet, but then started looking at a site on the Kenai Peninsula. In spring 1995, Kobe Steel announced it was pulling out, and the plant appeared dead unless Midrex finds other backers. The project called for importing iron oxide from South America and using Cook Inlet natural gas to fire the furnaces to produce iron pigs for export to steel mills in the Far East.

Titanium

The metallic element titanium is resistant to corrosion, and the metal and its alloys are extremely lightweight, yet strong. Titanium is used in aircrafts, spacecrafts, ships, guided missiles and armor plating for tanks. Titanium dioxide is used as a gemstone and for paint pigment.

– Adapted from *The Concise Columbia Encyclopedia*

Scrap dealer Bernie Karl inspects nickel-steel at his recycling facility in North Pole. Recycling helps the environment and, during times of high metal prices, can be profitable for Alaska's small businesses. Recycling has rebounded to levels of the late 1980s, when high metal prices and social concerns caused a boom for scrap dealers. Scrap metal is shipped from Alaska to buyers overseas and in the Lower 48; spent lead-acid batteries are shipped mostly to reprocessing plants in California. (Thomas K. Bundtzen)

LEAD

The principal ore mineral of lead is galena, a lead sulfide. Galena and other lead minerals are widely distributed in Alaska, but nearly all Alaska lead production has been as a by-product of polymetallic deposits mined for a more valuable target mineral.

In the past, a major source of lead was the old Alaska-Juneau Mine, which produced more than 40 million pounds of lead as a by-product. The lead content of the A-J ore was minor, but the galena could be conveniently recovered from the immense tonnage of gold ore put through the mill.

Most recently, Alaska's lead production jumped from essentially negligible amounts to third in the nation. Total lead production in Alaska since 1989 has been about 226,000 tons, because of Greens Creek and Red Dog mines where lead is a by-product of silver and zinc mining. Lead occurs in numerous deposits in Alaska associated with silver and zinc materials.

The 22 percent drop in lead prices in 1993 influenced the closure of Greens Creek. Even so, Red Dog's production of lead, while secondary in price and quantity to zinc, made Alaska third in the nation in mine-derived lead production in 1993.

MOLYBDENUM

Molybdenum occurs throughout Alaska. None of the deposits have been mined, although one came close.

The industrial mineral barite was once mined in Southeast. Powdered barite ore used as drilling mud in the oil industry was loaded onto ships from the mine site near Petersburg. (Steve McCutcheon)

That deposit, Quartz Hill, is one of the world's largest molybdenum reserves with up to 1.3 billion tons of minable ore. It is located 45 miles east of Ketchikan, inside Misty Fiords National Monument. The monument is designated wilderness closed to mining except for 156,210 acres around the deposit, which were exempted to allow for mine development.

U.S. Borax first explored the deposit in 1974 and planned to mine it as late as 1982. But work there ceased due to problems with developing a mine tailings disposal plan; the deposit was sold to Cominco Alaska in 1990.

ALASKA'S STRATEGIC MINERALS

During this century, Alaska has produced minerals considered "strategic" and "critical" to the nation's defense. These include tungsten, platinum, antimony, mercury, uranium and chromium.

Strategic minerals are primarily supplied by sources outside the nation's boundaries. Strict conservation measures have historically controlled the distribution of strategic minerals, especially during times of armed conflict or mineral embargoes.

Critical minerals are more easily obtained, because they can be either produced domestically or obtained in more adequate amounts from reliable foreign sources.

Since World War II, private industry and limited federally subsidized exploration have delineated economic or near-economic reserves of these 10 strategic minerals: cobalt, nickel, tin, platinum-group metals, mercury, fluorite, tungsten, antimony, chromite, uranium. Bauxite (aluminum), manganese, niobium, tantalum and mica exist in mineralized regions of the state, but no minable resources have been identified.

Lead

The biggest use of lead is in lead-acid batteries, accounting for 60 percent of the world's demand. Lead also is used to block harmful radiation from x-rays, television, video and computer monitors. When ingested, lead has been linked to serious health problems and its use in gasoline, pipes for drinking water and household paints has been phased out in the United States.

— Adapted from "Nonferrous Metals Outlook 1994, Lead," by John Keating, Natural Resources Canada.

URANIUM

The search for uranium in Alaska has correlated directly with the demand for uranium in peaceful and military energy applications. A flurry of prospecting occurred during the early- to mid-1950s when the Atomic Energy Commission stockpiled radioactive materials for nuclear weapons. Later during the energy crisis of the 1970s, investigations again intensified.

The only commercial uranium deposit found so far in Alaska was staked by amateur Ketchikan prospectors, Don Ross and

Molybdenum

Molybdenum, a hard, malleable substance with a high melting threshold, is used in X-ray and electronic tubes, electric furnaces, rocket and missile parts, and is a hardening agent in steel alloys.

The only molybdenum mineral of economic interest is molybdenite, the greasy, gray molybdenum sulfide. This mineral is used as a lubricant in spacecraft and automobiles.

— Adapted from *The Concise Columbia Encyclopedia* and *Alaska's Oil/Gas & Minerals Industry*

Reclamation

Cassiterite Placers was recognized by the state in 1994 for reclamation of its placer mine on Fox Creek near Fairbanks, shown here. Other mine operators also recognized were Alf Hopen for reclamation work on Little Eldorado Creek and John Brown for work on North Fork Harrison Creek. The Reclamation of the Year Award went to John McCain of The Mining Co. for outstanding reclamation of a mine on Ester Creek near Fairbanks. These were the first in an annual awards program by the state to recognize exemplary work in returning mined ground to useful condition. (Thomas K. Bundtzen)

By L.J. Campbell

Mining practices today in Alaska have changed considerably from the way they used to be done. One of the big differences is in the way miners restore their claims, returning the land and water to a condition equal to or better than it was when they started.

Before lifting the first spade of dirt or bulldozing open a pit on any claim of five acres or more, Alaska's miners must develop formal reclamation plans for each site. They envision what the site will be used for when mining ends, and then pattern their operation in such a way that the disturbed ground can be more easily put back into place during reclamation. They must file their reclamation plan with the state and obtain a reclamation permit before starting operations.

Reclamation became a legal requirement by Alaska statute in 1990, part of comprehensive legislation to clarify lease, rental and royalty payments for mining claims on state lands. Prior to this, miners had to identify their reclamation intentions to obtain state mining permits. The 1990 law formalized the process, extending reclamation standards to state, federal and private land, and established standards for closing mine sites. The law contained an additional requirement that miners post reclamation bonds up to $750 per acre. Reclamation costs typically run between $500 to $700 an acre.

Miners working less than five acres a year in one location are not required to obtain a reclamation permit or post bond, although they must still obtain all the other operational permits required by state and federal landowners.

A state bonding pool was established to help miners meet the bonding provision. Coming up with cash to post bonds was particularly difficult for small operators; banks and other traditional lending institutions are reluctant to finance this part of an operation. With the bonding pool, the miner pays a percentage of the required bond into an interest-bearing escrow account, and the interest funds the pool to cover reclamation costs should a miner default.

As of spring 1995, no miners have defaulted, according to Steve Borell, executive director of the Alaska Miners Association Inc. Any who do are charged five times the original bond amount on any subsequent mining attempt. "We as an industry put pressure (on miners) to get out there and clean up your mess," asserts Borell.

All mine sites must be restored to stable condition, but the extent of reclamation varies with different types of mining. Placers typically are surface mines with open cuts. Reclamation is relatively straightforward, in that the pit is filled, leveled, contoured to meet specified slopes, and seeded with native plants.

Reclamation of hardrock mine sites can be more problematic, however, largely because ore recovery methods are more complicated, often using chemicals to separate or leach out the minerals. This generates acidic wastes, both solid and liquid that must be treated. In this case, a mining company has to maintain waste treatment facilities long after the mine closes.

In planning for reclamation, the miner works with the landowner to determine the best post-mine use of the site and restores the land accordingly. The Fort Knox Mine site will someday become a recreation area with campgrounds and a lake, which state biologists will manage for fisheries. The mine construction, and its later operation, will be guided in part by the reclamation plan, implemented in stages through the mine's life.

The reclamation plan for Red Dog calls for restoring the site as wildlife habitat. The streams and lakes affected by mining operations will be returned to water quality adequate for fish. Red Dog Creek originally cut through the orebody and naturally held high mineral and metal levels; the company's treatment of the water to meet federal and state water quality standards should leave the creek in better condition that it was originally, according to Mitch Henning, manager of mine permitting and reclamation for the Alaska Division of Mining and Water Management. Mine operator Cominco Alaska will be responsible for maintaining treatment facilities on its tailings impoundment until the water meets state standards.

Elsewhere, reclamation has been underway at Cambior Alaska's Valdez Creek gold mine, scheduled to close in September 1995. The pit will be partially filled to create a lake suitable for recreation and fishing. Valdez Creek, which was diverted during mining, will be returned to its original channel. ■

After 11 years of open pit mining, Cambior Alaska closed its Valdez Creek Gold Mine in late 1995. Reclamation was ongoing during the mine's life. Placer gold was extracted from a series of pits, with each mined-out pit filled in with waste from the newly created pit above it. (Both, Thomas K. Bundtzen)

Uranium

This radioactive, hard and dense metallic element occurs naturally; the best known and most common uranium mineral is the black uranium oxide, pitchblende. Because uranium decays at a constant rate, the age of uranium samples can be estimated through radioactive dating. Uranium is used to fuel nuclear energy production and is a primary component in atomic bombs. The rare uranium-235 isotope is the only naturally occurring fission fuel for nuclear energy. Depleted uranium is widely used in military projectiles; its high specific gravity insures greater penetration into military targets. Peaceful applications include a variety of ballast and casing functions.

Kelly Adams, in the early 1950s at Bokan Mountain on Prince of Wales Island. Although small by most standards, these deposits are of high grade and easily accessible, because they are near water. About 120,000 tons of ore were mined by various operators from 1955 to 1971.

The Bokan deposit occurs as a breccialike accumulation of radioactive minerals and rare earth elements in a sodium-enriched granite. It was the first of its type to be recognized, prompting uranium geologists to look for Bokan-type radioactive occurrences worldwide.

Elsewhere in Alaska, uranium has been found in the westcentral part of the state, on the Seward Peninsula, in the Yukon-Tanana Uplands and in the Alaska Range.

Cobalt and Nickel

Cobalt is found combined with other metals in the ore minerals cobaltite and smaltite. Cobalt alloys are used in very hard cutting tools, high-strength permanent magnets and jet engines. Radioactive cobalt-60 is used in cancer therapy and to detect flaws in metal parts.

Nickel's chief ore minerals are garnierite, pentlandite and pyrrhotite. Nickel is present in most meteorites, and trace amounts are found in plants and animals. Its primary use is in alloys, for its strength, ductility and resistance to corrosion and heat. Many stainless steels contain nickel.

— From *The Concise Columbia Encyclopedia*

COBALT AND NICKEL

As of yet, there has never been any production of cobalt or nickel in Alaska.

Resources of cobalt metal, however, are contained in four deposits in Southeast, Brady Glacier, Yakobi Island, Mirror Harbor and Funter Bay. These deposits amount to about 63 million pounds, or about three years of annual cobalt consumption in the United States. They also contain an estimated 1.12 billion pounds of proven reserves of nickel, an amount equal to about two years current U.S. consumption. The bulk of cobalt and nickel resources and reserves are contained in the large Brady Glacier sulfide deposit in Glacier Bay National Park and Preserve. This deposit contains from 24 million to 54 million pounds of cobalt, but it has not been studied to determine the success of cobalt recovery.

Another potentially significant cobalt resource exists at the Bornite copper deposit.

Bokan Mountain holds Alaska's only known commercial uranium deposit. (Staff)

PLATINUM

In past years, Alaska held the title of largest producer of platinum metals in the nation. This changed with the initiation of platinum metal production from the Stillwater complex in Montana several years ago.

Almost all of Alaska's total production of refined platinum, 575,000 ounces, came from two regions — the Goodnews Bay District in southwestern Alaska and the Salt Chuck copper-platinum mine on the Kasaan Peninsula.

Platinum at Goodnews Bay in the Salmon River drainage was discovered in 1926, 25 years after gold discoveries in the region, by an Eskimo prospector named Walter Smith. Suspecting he might have stumbled upon a significant mineral, he sent a sample of a steel-gray metal from his claim to the territorial Department of Mines, which identified it as platinum. By 1937, large-scale dredging operations were underway in the Salmon River drainage, and through 1981, 545,000 ounces of refined platinum metals had been extracted.

The hardrock platinum deposits at Salt Chuck were originally mined for copper, but in 1918 samples were run for platinum metals and nickel. Platinum and a related metal, palladium, turned up. By 1941, 19,000 ounces of palladium and 2,500 ounces of platinum had been produced as a by-product of copper mining.

This dredge mined platinum at Goodnews Bay in the Salmon River drainage, part of large-scale platinum mining that occurred there from 1937 to 1981. (Arne Bakke)

Platinum

More than 668,540 ounces of elements from the platinum group have come from Alaska mines through 1993. Platinum group elements, or PGEs, include platinum, palladium, iridium, rhodium (extremely rare), osmiridium, osmium, rhenium and ruthenium. PGEs are used in surgical tools, laboratory utensils, electrical-resistance wires, contact points, jewelry, dentistry, powerful magnets and in automobile catalytic converters. Palladium occurs in greater abundance than other PGEs and is becoming increasingly important in the manufacture of catalytic converters as a substitute for the more costly platinum. This pan of "platinum" came from the Goodnews Bay Mine. The mine's production contained about 85 percent platinum, 4 percent iridium, 3 percent palladium, 2 percent osmium and 6 percent other PGEs.

— From *The Concise Columbia Encyclopedia*
(Photo by Steve McCutcheon)

Several thousand ounces of platinum metals have been recovered as by-products of placer gold mining since 1902, most notably from the Cache Creek, Chistochina and eastern Seward Peninsula areas.

The largest resources of platinum metals remaining in Alaska, about 1 million ounces, are found at Goodnews Bay and in the Brady Glacier nickel-copper deposits. The platinum reserves at Goodnews Bay are mainly contained in deep placer deposits; however, exploration for more shallow placer and hardrock platinum deposits has been underway and may lead to development and modest production there in the near future.

Metallurgical research conducted by the U.S. Geological Survey, in cooperation with Newmont Mining Corp., discovered a resource of about 580,000 ounces of platinum metals in the Brady Glacier deposits. However, it's unknown if the platinum would be recoverable should mining ever occur on this property.

Despite significant past production, platinum metals have been Alaska's sleepers. Most important deposits have been discovered by accident; new occurrences continue turning up. Prime areas of future prospecting will be mineral belts in southwestern Alaska and exploration of igneous complexes in Southeast.

CHROMIUM

Chrome mineralization is found in a variety of deposits in at least six major mineral

Chromium

This silver-gray, shiny metal occurs only in compounds. Its chief source is the mineral chromite. Hard and nontarnishing, chromium is used to plate other metals and to harden steel. It contributes hardness, strength and heat resistance when used in alloys. Chromium compounds are used as paint pigments, in tanning and in dyeing.

— From *The Concise Columbia Encyclopedia*

belts in Alaska, but has only been mined on the Kenai Peninsula where production was largely subsidized by the federal government.

When critical shortages occurred during World War I, about 2,000 tons of metallurgical-grade chrome was produced from the Claim Point Mine near Seldovia. During World War II and the Korean War, chrome shortages prompted production from deposits at Red Mountain, about 12 miles east of Claim Point. Between 1943 and 1957, about 28,829 tons of ore were produced from orebodies within Red Mountain; the mining was subsidized by a federal price support program which expired in 1958. Most United States domestic production ended in 1961.

Red garnets found near Wrangell are popular collector's specimens. Rarely are Alaska's garnets of jewelry quality, because of imperfections making them difficult to cut. (Steve McCutcheon)

Based on work by J.Y. Foley and others in the U.S. Bureau of Mines, reserves statewide contain about 4.3 million tons of chromic oxide, half of which are found in deposits on the Kenai Peninsula. In particular, the Red Mountain deposits south of Seldovia may represent one of the largest low-grade chrome resources in the nation.

Other chromite belts in Alaska include deposits on the Kanuti River, a tributary of the Koyukuk River; Eklutna near Anchorage; in the mountains near Chitina; eastern Chichagof Island; and in the DeLong Mountains. Because chrome is a low unit value ore, only those deposits near transportation systems are considered to have economic potential.

Tin

This metallic element has been used by humans for centuries. The chief ore mineral is cassiterite, or tin stone. This very soft metal can be rolled, pressed or hammered into extremely thin sheets, such as tin foil. A tin coating, applied by dipping or electroplating, protects iron, steel, copper and other metals from rust. Compounds of tin are used for dyeing, weighing silk and as reducing agents. Stannous fluoride, added to toothpastes and water supplies, prevents tooth decay. Toxic tin organic compounds are used as fungicides and catalysts.

— From *The Concise Columbia Encyclopedia.*

TIN

Although total state production of tin, about 7.3 million pounds through 1993, has been modest compared to other world sources, Alaska's production still constitutes the nation's largest primary source of the metal.

The strategic importance of tin in Alaska has long been known. As early as 1902, placer tin deposits on the western Seward Peninsula were successfully mined. Both lode and placer deposits of the Lost River near Teller became the focus of national concern. By late 1942, two-thirds of the tin resources and three-fourths of the world's tin smelter capacity had been taken over by Japanese occupation of Malaysia. The War Production Board immediately approved government construction of a 500-ton-per-day mill at Lost River, but the project depended on support from the Navy to keep open shipping. Because of Japanese occupation in the Aleutian Islands, U.S. military leaders decided not to divert naval resources and the mill construction was canceled. The government did continue exploring the deposit, however, and by the end of the war had delineated a significant tonnage of high-grade tin ore.

The onset of the Korean War rejuvenated the mill project. By 1956, 51,000 tons of ore averaging 1.13 percent tin had been taken. After the Korean War, the mine closed.

Extensive exploration in the area continued and by the mid-1970s had blocked out an inferred reserve of 124 million pounds, one of the largest single reserves of tin in North America. Other important Alaska tin deposits include the Kougarok lode on the Seward Peninsula, the Sleitat lode northwest of Iliamna, and Coal Creek north of Talkeetna.

The other major area in Alaska for tin production was the Manley District north of Fairbanks where more than 700,000 pounds of cassiterite were recovered as a by-product of gold mining. From 1979 to 1990, Alaska produced about 120,000 pounds of tin concentrate annually from small placer operations on the western Seward Peninsula and the Manley area – the only primary source of tin in the United States. Since 1990, tin production has been restricted to the Manley and Ruby districts, all as by-products of gold mining.

TUNGSTEN

Modest amounts of high-grade tungsten concentrate were shipped from Alaska during both world wars and the Korean War. Early Fairbanks-area gold prospectors found tungsten deposits, and demand during World War I led to some production from small deposits on Gilmore Dome.

The same pattern emerged during World War II in the Hyder District of Southeast. A former gold-silver metal mine, the Riverside Mine, reopened to ship tungsten concentrates during the Korean War.

During the Korean War, a federal program subsidized exploration and production of tungsten throughout Alaska, resulting in tungsten mining on the western Seward Peninsula, in Southeast and the Fairbanks area. Total tungsten output, about 286,000

An old mining cabin near Hyder in Southeast overlooks Salmon Glacier. The Hyder District produced high-grade tungsten-copper-lead-zinc-silver ore from 1925 to 1951, as well as some placer gold. Glaciers created most of Alaska's placer gold deposits as they moved over hardrock lodes, grinding them down and carrying away the pieces. (Patrick Windsor)

Tungsten

Tungsten, a silver-white to steel-gray metallic element, is one of the densest, hardest metals known with the highest melting point of any metal. It's used for wires and filaments in light bulbs and electronic tubes. Tungsten carbide is used in place of diamonds as an abrasive.
— From *The Concise Columbia Encyclopedia*

pounds, pales in comparison to national requirements, but significant resources have been delineated at Lost River tin deposits, a lode near Fairbanks and other scattered sites.

ANTIMONY

Antimony is plentiful in Alaska and is often associated with gold mineralization. Antimony ores from at least 16 deposits were shipped to markets at various times through the 1970s, usually in relation to high demands for war and related industries. Production was subsidized by the federal government during the Korean War. Alaska's total yield of antimony, about 11.1 million pounds, amounts to about 7 percent of domestic production through 1990.

During wartime, Alaska's contribution figured more importantly. For instance from 1936 to 1942, the Stampede Mine in the Kantishna District is credited with about three-fourths of U.S. domestic production during the war. Although large resources remain in the Fairbanks and Kantishna areas, antimony mines in Idaho, Montana and Nevada now overshadow Alaska's contribution.

Antimony

Antimony is a semimetallic metal, silvery blue-white, brittle, easily powdered and a poor conductor of heat and electricity. Its chief uses are as alloys and compounds in storage batteries, cable sheathing and paint pigments.
— From *The Concise Columbia Encyclopedia*

Sand, gravel and building stone are important industrial minerals in Alaska, worth almost $67.5 million in 1994. Here trucks work at the Cape Nome quarry, operated by Sitnasuak Corp. and Bering Straits Native Corp. The quarry produces rip rap and gravel for roads, construction projects and sea walls, and it has been used as a nesting site for peregrine falcons. (George Matz)

MERCURY

A large mercury belt in southwestern Alaska extends from Marsh Mountain near Dillingham to the Cripple Mountains north of McGrath. A dozen mines have recovered more than 40,000 76-pound flasks of mercury intermittently since World War I. The principal producer was the Red Devil Mine near Sleetmute, which operated off and on between 1942 and 1972 with continuous mining from 1954 to 1963.

During high prices of the 1950s, Alaska mercury mines supplied the nation with 10 percent to 20 percent of its mercury requirements.

Mercury was classified strategic because of its use in detonators and control instrumentation; some of the mercury production in Alaska was subsidized by the federal government.

OTHER STRATEGIC MINERALS

Fluorite is often found in tin deposits statewide. The only viable Alaska reserve is within deposits at Lost River, previously described for tin and tungsten content. According to an estimate in the 1970s, these deposits contain about 25 percent of the nation's total reserve base of fluorite. This mineral is used in steel manufacture as a fluxing agent.

Niobium, tantalum, manganese and titanium all occur in Alaska, but no reserves of any of these have been published. Small quantities of niobium and tantalum exist in tin deposits, and could be a by-product should tin mining commence. Small lodes of manganese exist in the Alaska Range and Yukon-Tanana Uplands. Titanium occurs in enormous low-grade lode iron deposits and in some rich beach placers in Southeast. The U.S. Bureau of Mines recently estimated that beach placer deposits contain about 150 million tons of titanium-bearing beach sands in two deposits north of Brady Glacier.

INDUSTRIAL MINERALS

Jade, soapstone, sand, gravel, chemical and structural-grade limestone, gypsum, garnets, graphite, asbestos, barite, pumice, clay and marble are among Alaska's industrial minerals. All have been mined here at one time or another.

Industrial minerals contributed more than $67.5 million to Alaska's total mineral value in 1994. Sand and gravel production of about 13.9 million tons worth nearly $42 million accounted for the bulk of that. About 640 people work full-time in sand and gravel quarries and production.

SAND AND GRAVEL

You don't have to be a geologist or a miner to find sand and gravel in Alaska. Unlike gold and silver, sand and gravel seem to be everywhere. In an attempt to keep the grit outside, Alaskans honor the unspoken custom of removing their shoes at the door when they go inside each other's homes.

But road builders, port developers and structural engineers like Alaska's ready supply. Gravel is one of the state's most important commodities. Since 1948, more than 1.1 billion tons of aggregate have been mined.

Construction projects, particularly roads, keep demand high for gravel. In Anchorage, a chronic shortage of gravel created by the city's growth has necessitated importation of gravel from the Matanuska River valley. Gravel mine tailings from gold dredging and modern flood plain deposits are important local sources in the Fairbanks area. River deposits on the North Slope supply rock for development of northern oil fields.

In 1994, Sitnasuak Corp. and Bering Straits Native Corp. continued to quarry rip rap at the Cape Nome location. These operations have increased in importance in recent years. In 1994, Sound Quarry Inc., jointly owned by the corporations, mined about 100,000 tons of high-quality rip rap for sea wall construction projects at Shismaref, Point Hope and Bethel.

The state Department of Transportation and Alyeska Pipeline Service Co. mined and quarried significant amounts of stone and aggregate for maintenance of roads and the trans-Alaska pipeline. Calista Corp. produced record levels of sand and gravel for airport and road construction through its region, and Sealaska Corp. sold large amounts of shot rock (12-inch to 36-inch natural rocks) for construction projects throughout Southeast.

The growing city of Anchorage has increasingly fewer sources of minable gravel and depends on shipments from quarries in the Matanuska Valley. Gravel quarried south of Palmer is conveyed to this stockpile and loading facility on the Glenn Highway, where gravel is dumped into Alaska Railroad hoppers for delivery to Anchorage. (Steve McCutcheon)

MARBLE AND BUILDING STONE

Long before the Russians arrived, the Tlingit Indians of Southeast carved marble for utensils, art and religious objects. Later, the marble and limestone on Prince of Wales Island were among the first mineral resources mentioned in U.S. Geological Survey reports of the late 1800s. In the 1890s,

Mercury

Mercury, or quicksilver, is the only common metal existing as a liquid at ordinary temperatures. This silver-white, mirrorlike metal is used in thermometers, electric switches, mercury vapor lamps, and certain batteries.

Mercury vapor and certain mercury compounds – particularly the organic ones – are toxic. Although common mercury sulfides, such as cinnabar, are relatively stable, any person using mercury should be aware of the potential hazard. It isn't easily discharged from the body and can accumulate to toxic levels causing skin disorders, liver and kidney damage, and other health problems. Concerns about mercury pollution from industrial discharges led to an international ban on dumping mercury in the oceans to protect the food web, signed by 90 nations in 1972. Mercury's use in many pesticides and other industrial products has been prohibited in the United States.

Despite the undesirable effects of mercury pollution, its safe use in batteries, laboratories and control instruments makes substitution impractical.

— Adapted from *The Concise Columbia Encyclopedia* and *Alaska's Oil/Gas & Minerals* marble from Ham Island near Wrangell was worked by Natives for tombstones, to replace wooden totems as they adopted burial customs introduced by westerners.

Urban growth on the West Coast around 1900 created a demand for building stone and shipments of high quality marble commenced from quarries at Tokeen. By 1949, at least 2.15 million tons of chemical-grade stone and 450,000 tons of structural-grade limestone were quarried from a dozen sites on Prince of Wales and Dall islands, although the principal workings remained in the Tokeen area.

Through 1920, more than 72 buildings in Washington, Oregon, California, Idaho, Utah, Montana, Minnesota, Massachusetts and Pennsylvania used Alaska marble for interior work. The state Capitol in Juneau was built in 1930 with four exterior columns of Tokeen marble, plus marble wainscoting and trim inside.

Changes in tastes and building styles, inflation, and development of marble quarries in western states helped end Southeast's marble industry in the 1940s. Still, measured reserves of about 800 million tons are available in Southeast for future development. Sealaska Corp. has been actively exploring high-grade limestone deposits on Prince of Wales and Dall islands for export.

Chemical-grade limestone deposits suitable for concrete have been drilled along the Alaska Railroad near Cantwell. Several hundred million tons of reserves are known from three deposits. The new Healy Clean Coal Project will require limestone in its flue process and local quarries have been sought.

A small limestone quarry has produced a crushed limestone product for the past couple of years. Alaska Limestone Inc. operates a small quarry and a crushing and bagging plant about four miles north of Cantwell.

Basalt is being mined near Fairbanks at the Yutan Construction Co. basalt quarry on Badger Road. Several million tons of high quality basalt have been produced for crushed rock in septic leach fields, as fill for diversion dams, as asphalt road metal, and for uses as ornamental and building stone.

JADE

Much of Alaska's jade has come from the Jade Mountains of northwestern Alaska.

Early explorers found Natives using jade stone as ornaments and for cutting tools. Early this century, miners tried to exploit the jade, but the large size of the boulders complicated their task and made it generally unprofitable because transportation to commercial centers was so difficult.

In 1952, Archie Ferguson of Kotzebue developed a portable wire saw to slice the boulders in the field, and this did much to solve the field-to-market problem.

NANA, the Native regional corporation, purchased most of the Jade Mountains claims, starting with Gene Joiner's property in 1976. The corporation built a modern shop in Kotzebue for cutting and polishing the stone, marketed under Jade Mountain Products Inc.

Today, NANA's holdings include significant deposits in the Jade Mountains. NANA also purchased the old Stewart Jewel Jade Mine on Dahl Creek in the Kobuk area. That former gold mining claim was purchased by Ivan and Oro Stewart in 1966 and operated during the summers, accommodating occasional groups of tourists, until Ivan died in 1986. Jade from that site will be used as interior tiles in the lobby of the new Alaska Native Hospital in Anchorage, according to Mrs. Stewart.

Other examples of jade from the Kobuk River valley used for ornamentation can be seen at the National Guard complex north of Anchorage.

Alaska's jade is known internationally. In the 1950s, Argentina's president Juan Peron purchased one of the largest jade nuggets ever found in the Kobuk region. He bought this 12-metric-ton jade boulder to become a memorial to his late wife, Eva. But by the time the jade was ready, Peron was out of power; the nugget was eventually cut up and marketed in smaller pieces.

Two kinds of minerals — jadeite and nephrite — are commonly referred to as jade. Only nephrite is found in Alaska. This boulder from Dahl Creek has been sawed open, revealing the green jade. (Steve McCutcheon)

OTHER INDUSTRIAL MINERALS

Alaska has produced several other industrial minerals in the past including barite, gypsum, asbestos, pottery and brick-quality clays, garnets, and small amounts of pumice.

Barite can occur as a white, yellow, blue,

red or colorless sulfate mineral used in manufacturing linoleum, oil cloth, rubber and plastics. The biggest use of barite — 85 percent of world consumption — is in oil production as mud to seal the wells during drilling. Some deposits in Southeast have been exploited: a 1915 shipment of barite from Prince of Wales Island and extraction from a lode near Petersburg from 1967 through 1980. Most of that barite was mined offshore, and a small amount was bagged as drilling mud for Alaska's oil fields.

Clay deposits in Alaska currently are tapped only by artists and production potters. They use Bootlegger Cove clay from Cook Inlet and clay from the Usibelli Coal Mine at Healy. A University of Alaska researcher studied producing bricks, ceramic tile, sewer pipe and flue tile from clay in the Healy coal fields, and the feasibility of developing a clay industry seemed to hinge on demand. Another researcher looked at creating high quality porcelain from kaolinite-based clays from Interior Alaska sericitic, or potasium-bearing mica, schist. Porcelain containers have been made on a limited basis from these clays near Tenderfoot in the Interior.

For a short time after World War II, Anchorage had a brick plant that used clay from Bootlegger Cove and Sheep Mountain on the Glenn Highway. The plant and kiln operated until 1948.

From 1902 until 1926, gypsum was mined on eastern Chichagof Island. Although Alaska has known gypsum deposits, such as those that add to the color of the hillsides in the Sheep Mountain area along the Glenn Highway, none of those deposits are currently being developed because of uncertain economics. Gypsum is used in sheet rock and cement, and demand fluctuates with the housing industry.

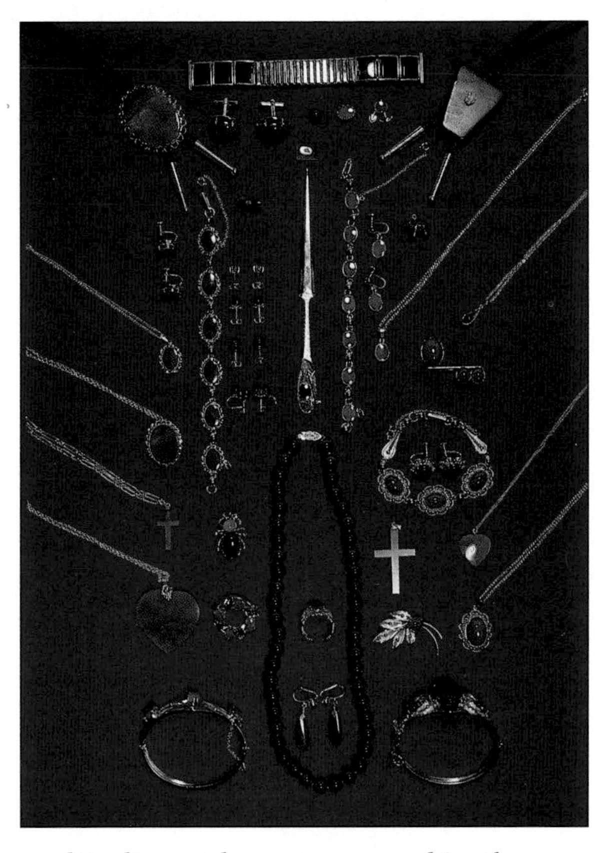

Jewelry made with Alaska gold and jade, as well as jade and soapstone carvings, have been popular souvenir and collector's items for many years. Now other Alaska minerals are finding a place on the counter as well. A pink zirconium mineral called euydialyte, extracted in small quantities here, is being used in jewelry. Alaska also has gem-quality rhodonite and aquamarine. While the iron mineral hematite, marketed as Alaska Black Diamond, and the copper minerals azurite and malachite, marketed as Arctic Opals, are found here, much of the jewelry sold as such is made with imported stones. (Steve McCutcheon)

During World War II, the federal government declared asbestos a strategic mineral and urged development of domestic reserves. Selected asbestos ores from the Kobuk Valley were freighted along winter trails to the coast for shipment to a buyer in 1943, but no larger development took place.

In 1980, Doyon Corp. announced discovery of a major asbestos deposit at Slate Creek southwest of Eagle. The company's drilling showed it to be one of the country's largest deposits, and it was considered to be an important North American find. But health problems related to unwise asbestos use has severely curtailed domestic markets and the Slate Creek deposit remains undeveloped.

From 1912 to 1920, high quality almandine garnets were mined from schist and migmatite near Wrangell. They were collected as museum specimens and used in abrasive applications, because of their hardness and other properties. A rock ledge containing garnets was deeded to the children of Wrangell in 1962, and they would gather the stones to sell to tourists visiting town off the ferries and cruise ships. The ledge was closed down in the early 1990s.

In the early- to mid-1980s, a few gem-quality diamonds were recovered in placer gold operations in the Upper Chena River and Birch Creek drainages, spurring sporadic exploration. More recent exploration has attempted to find the hardrock source of these placer diamonds, focusing on geologic trends similar to those associated with significant diamond discoveries in the Northwest Territories of Canada.

Zeolites, perlite, diatomaceous earth, sulfur and pumice have been examined for production potential in Southcentral and on the Alaska Peninsula. Only small amounts of pumice have been extracted for use in the Anchorage area. ◗

Fort Knox Mine

By Richard Montagna

Editor's note: *Richard Montagna is a free-lance photographer and writer from McGrath whose work has appeared in other* ALASKA GEOGRAPHIC® *quarterlies. The author would like to thank Richard Flanders, former project geologist with Nixon Fork Mining, and Chris Puchner, vice president of exploration with Nixon Fork Mining, for their help with this article.*

Located on 7,000 acres 15 miles northeast of Fairbanks in the Fish Creek drainage, the Fort Knox Mine being developed by Fairbanks Gold Mining Inc. will be the largest open pit mine in Alaska, and the largest mining project in the Fairbanks area since the days when dredges worked nearby creeks. Construction work on the gold mine began in spring 1995 and start-up is projected for October 1996.

The Fort Knox Mine is being developed in an area with a long history of mining activity. Felix Pedro, the prospector credited with discovering gold in the Fairbanks area, first staked a mining claim on Fish Creek in 1902. Other early Fairbanks prospectors, such as Frank Cleary and Tom Gilmore, soon followed and by 1907, 11 cabins were located on the road by upper Fish Creek.

The miners who worked the drainage employed a variety of methods for removing the gold. Nels Jackson began drift mining before switching to open cut hydraulic mining. Bob Health and his crew ran a dragline and bulldozer. In all, 10 or so independent miners operated small-scale placer operations in the drainage.

Besides the placer claims, there were several hardrock, or lode, claims being worked in the drainage. Ore-bearing rock was crushed in a mill to release the gold and to concentrate the ore for shipping. In 1913, several hard-rock claims were in operation: Edward Voght mined a gold-quartz vein along the ridge between Melba and Monte Carlo creeks, Mike Stepovich and Charles Lockhart had a tunnel at the head of Fish Creek and prospectors Murphy and Perrault had a claim on upper Pearl Creek that was reported to produce $30 of gold per ton of ore. From 1913 to 1915, a three-stamp mill that had been built in Fairbanks in 1909 was operated on Melba Creek by Edward Voght.

Construction crews clear trees from the site of the Fort Knox Mine northeast of Fairbanks. When fully operational, Fort Knox will be the largest open pit mine in the state. (Richard Montagna)

Small-scale miners were not the only ones interested in the Fish Creek area. In 1925, the Tanana Valley Gold Dredging Co. built a camp, hydraulic ditch and the Tanana Valley Dredge on the creek. They ran the dredge intermittently from 1926 to 1935, when the property went into receivership to the Northern Commercial Co.

In 1937, the Fairbanks Exploration Co. (F.E. Co.) purchased the Tanana Valley Gold Dredging property, its dozen buildings and the dredge. They renamed the dredge Dredge No. 7, ran a power line to Fairbanks and built more buildings. The dredge operated in 1940 and 1941 and took out gold from a two-mile stretch of the creek. However, in October 1942 the government shut down all non-strategic mining, including gold mining, to boost the war effort.

In the years following World War II, several operations picked up again on Fish Creek, but not Dredge No. 7. In 1959, Wolfe Creek Mining Co. moved into Fish Creek. In 1964, Walter Roman purchased Wolfe Creek's claims and leased others for a placer operation on Pearl Creek that his family has mined for more than 30 years. Roman dug a mile-long ditch south of Fish Creek that redirected water to his dragline and hydraulic operation on Pearl Creek.

The current Fort Knox project had its beginnings in 1980 when 19 placer claims were staked on Monte Carlo and Melba creeks by Joe Taylor, George Johnson and

Workers use cables to assemble steel girders along Fish Creek. (Richard Montagna)

two other local prospectors. During the next few years, they found they could pan gold all over the ridge between the two creeks and staked an additional 34 claims, covering a granite outcrop that had been mapped in the early 1900s. Taylor and Johnson eventually acquired full ownership of the claims and, as annual assessment work revealed the presence of visible gold in the granite, the deposit's importance increased.

Serious exploration of the deposit began in 1987. Between then and 1991, several companies were involved in trenching, soil sampling, environmental studies and drilling more than 350 test holes. Fairbanks Gold Ltd. and their partner, Ventures Trident Ltd., invested more than $17.5 million in determining the extent of the deposit. Amax Gold Inc., a large aluminum and energy-based company, took control of the project in 1992, after purchasing the outstanding stock of Fairbanks Gold Ltd. and its partners. Fairbanks Gold Mining Inc. is a subsidiary of Amax.

Fort Knox will be the largest gold mine in the country using a mechanical ore processing system. Development of the mine will include a 120- to 150-acre, open pit mine on Gilmore Dome, the mill site, roads, water supply and lines, a tailings dam and impoundment area and a water reservoir. The tailings dam will exceed 300 feet in height, extend 4,000 feet across the valley and be used to hold water for processing operations. After the project closes, the open pit will slowly fill with water, leaving a 400- to 600-foot-deep reservoir.

The mine has an estimated life of 16 years, will operate 24 hours a day, 365 days a year, and is expected to produce 4.1 million ounces of gold. The gold-bearing ore will be crushed in a mill that will process 35,000 to 50,000 tons of ore per day and produce 300,000 to 350,000 ounces of gold annually. Small amounts of bismuth are found in the ore and might be recovered as a by-product. Large dump trucks will transport the ore from the pit to the mill site, where it will be crushed and ground. The material will then be fed into cyanide leach tanks where the gold goes into solution. Granular-activated carbon particles will absorb the dissolved gold, which continues through several more processes before being shipped out to a refiner. Gold recovery is estimated at 91 percent.

To get approval to developing the mine, which is on lands owned and managed by the state, Fairbanks Gold Mining Inc. underwent a series of meetings and studies during a three-year period. According to Dick LeFebvre, deputy director of the Division of Lands, and project manager for Fort Knox, one of the things his agency looked at was the condition the area would be in when the mine was finished. "The area had been heavily disturbed by previous mining activity and we wanted to leave a legacy that the area would be in better shape than when we found it. The project will leave a recreation area with a reservoir, improved wildlife habitat, wetlands complex and fisheries."

For most people in the Fairbanks area, their concern focused on the economic impact of the project. Design and construction costs are estimated at $255 million. The project should employ several hundred seasonal construction workers. Long-term economic impact is estimated at $60 to $70 million per year, including employment of 200 to 275 year-round, and expenditures for support services and power consumption.

Since gold deposits often occur in clusters, discovery of the Fort Knox deposit led to renewed interest in mineral exploration of the area. Other large, low-grade lode deposits were identified and staked in the Fort Knox vicinity even before construction of the project began. With other similar projects being looked at, Fort Knox may be the start of a new chapter in the history of the Fairbanks Mining District. ◗

RIGHT: *One-ton bags of rocks, collected from bulk samples excavated in 1990, await processing. The bulk sampling is done to help test the reliability of drilling methods. (Arne Bakke)*

BELOW: *Drill cores of granite are prepared for logging by geologists. (Arne Bakke)*

JOHN DEERE

Nixon Fork Mine

By Richard Montagna

In 1917, prospector R.D. Mathews discovered placer gold on Hidden Creek, approximately 35 miles northeast of McGrath. Subsequent discoveries were soon made on Mystery, Ruby and Submarine creeks. As the claims were worked, the miners discovered that the gold became more abundant as the deposits were followed up the creeks but that above certain points gold was not found. Shafts sunk at these points revealed rich gold-bearing ore. In 1918, Pearson and Strand staked the Crystal Lode and by the following year other lode discoveries had been made: the Whalen Lode at the head of Holmes Gulch by Whalen and Griffen, a lode on Mystery Creek by Mespelt and McGowan and a number of other claims. These discoveries led to the development of what is now known as the Nixon Fork Mine, the only commercial hardrock gold mine in the Kuskokwim River drainage.

From the beginning of its operation in 1917 until its closure in the early 1960s, the Nixon Fork Mine produced approximately 59,500 ounces of lode gold and an estimated 15,000 ounces in placer gold from adjacent creeks and hills.

In the first year of discovery, the Pearson-Strand property was optioned out for $24,000 to T.P. Aitken, who developed the Crystal shaft and began operation. By the end of 1919, 400 tons of ore, averaging $80 in gold per ton, had been shipped out. Aitken, however, made only $5,000 in payments and, still owing $19,000, he forfeited his lease agreement within a year.

Arnold "Bo" Demientieff leaves McGrath with a load of freight for the Nixon Fork Mine. (Richard Montagna)

In 1920, the Alaska Treadwell Gold Mining Co. (also called the Treadwell Yukon Co., or just Treadwell) became interested in the property and took out options on several of the claims. Asking price for the claims was high: $500,000 for the Whalen claims to be paid over five years, $350,000 for the Pearson-Strand claims over three years and $325,000 for the Mespelt claims over four years. Treadwell made an initial payment of $5,000 for an option on the Whalen properties, $10,000 for an option on the Pearson-Strand property and $2,000 on the Mespelt property. The Mespelt option was soon dropped and the other contracts renegotiated to pay a percentage of the profits. Treadwell also sank several shafts of its own and built a 10-stamp mill for crushing the ore.

Built in 1921, the 10-stamp mill was the center of operations. Here the ore was trammed to the top of the mill where it first went through a grizzly, a large screening device that separated the rock by size. Rock small enough to fit through the openings passed on to the jaw crusher, while the larger rocks were either discarded or broken with sledge hammers. Like a large mechanical vise with heavy steel plates, the jaw crusher would separate and smash the rock into 1/2- to 1-inch chunks. The material then went through the 10-stamp mill where the pistonlike action of the stamps reduced the rock even further. From the stamp mill the material was sent to a classifier where the fines were run off and the coarser material was conveyed to a ball mill for further crushing. The flow from the ball mill was run through an amalgamator, mixed with mercury, then passed over large copper plates where the fine gold was trapped. The mill was powered by a 125-horsepower

FAR LEFT: *Longtime McGrath resident Willard "Tex" Gates (1909-1993) was born in Fulchar, Texas. He left there during the Depression to ride the rails looking for work and ended up in Alaska in 1936. He first worked in Fairbanks and Tanana cutting wood before coming to McGrath in 1940, then on to Ophir to work as a miner. Gates worked for Eric Hard on Bear Creek, eventually buying the claim in 1959 and working it until 1985. Gates said when he worked for wages, he worked five months of the year, seven days a week, 10 hours a day for $6 to $7 a day. During World War II, he was a combat engineer in the Aleutians and a member of the Alaska Scouts. He died at his home in McGrath in 1993. (Richard Montagna)*

LEFT: *Richard Flanders, project geologist with Central Alaska Gold Co., examines some of more than 30,000 feet of core samples drilled between 1990 and 1992. (Richard Montagna)*

steam engine that ran off of two 70-horsepower boilers. The boilers reportedly used 300 cords of wood a year, cut by one individual. The mill had a capacity of 50 tons of ore per day and in the first four months of 1922 produced a reported $114,024 in gold. Gold embedded in sulfide ore was not removable by crushing and gravity separation and had to be shipped out for processing to a smelter in Tacoma, Wash.

Although the ore in some of the claims was high quality, the orebodies were considered small. For example, the Whalen orebody was estimated at 3,000 tons of ore valued at $63 of gold per ton. Because of this, and the high purchase price of the claims, Treadwell did not consider the mine as having potential for long-term development. Treadwell planned to expose the ores as rapidly as possible, and forfeit its options to buy as soon as quantities of ore diminished.

Costs for the operation, by 1920 standards, were high. Labor was based on $6 per day, board at $3 per day, wood to fire the boilers was $9 a cord, lumber was $60 per 1,000 board feet and freight was $80 per ton from Seattle to Berry's Landing, now called Medfra, and, depending on the season, another $20 to $80 per ton on the 12-mile road to the mine. In all, almost $240,000 had been spent by the end of 1920, even before the mill was built and operating.

Treadwell ran the mill for two years, taking out a reported $235,000 in gold. Finally in 1923, with the occurrence and size of ore shoots decreasing, it dropped its options on the claims. The mill was subsequently leased in 1924 by E.M. Whalen, who is reported to have earned $80,000 in gold from milling the broken ore remaining in the Whalen shaft. Pearson and Strand took over the mill and the relinquished claims, but the property soon passed to the Mespelt brothers in 1926.

Unlike most of the operations in the Kuskokwim region that removed the gold by placer mining, the gold lode was removed from the Nixon Fork Mine through a series of 11 shallow shafts of varying depths. Most of the gold Treadwell recovered came from the Whalen shaft that had been developed to 110 feet. Crystal shaft was sunk to 65 feet and Garnet shaft eventually reached 460 feet. The underground workings totaled about 2,000 feet.

The mine operated intermittently for another 40 years with the Mespelts working the lodes continuously from 1926 to 1933, and intermittently until leasing it to H.G. Wilcox in 1950-1951. The property was then leased to Strandberg and Sons who held the leases until the early 1960s. Most reports estimate that about $1,315,000 in gold was taken out of the Nixon Fork throughout the years.

In 1963 Ted Almsy and Margaret Mespelt of McGrath began restaking claims in the area. The lease to the Standbergs had failed to include a reversion clause and Margaret wanted to protect the Mespelt claims. Ted thought the property had potential for another type of mining: "I wasn't interested in getting into the gold mining business. I was involved in the tourist business as a registered guide. But all the USGS reports I read mentioned copper a lot and I thought maybe I could get a copper company interested in it and use the money for my tourist business. Gold mining wasn't economical. The cost of mining had gone up six times from what it had been 20 years before, but gold was still only $35 an ounce."

Ted contacted many of the largest mining companies in the world who looked at the property for both copper and gold, but it took 20 years before any serious exploration began.

In 1983, Battle Mountain Gold Corp. began exploration on the property now held by the Mespelt and Almsy Mining Co. of McGrath. During the next five years, it

Heavy snowfall in the early 1990s collapsed the roof of the 10-stamp mill used to process ore during earlier operations of the Nixon Fork Mine. (Richard Montagna)

drilled 24,000 of exploratory holes averaging 283 feet per hole.

In 1989, Central Alaska Gold and Associates entered into a joint venture with Battle Mountain to continue exploration. The following year they drilled 30,000 feet with promising results, and returned in 1992 to drill another 1,400 feet.

Final exploration to determine the economic viability of the Nixon Fork Mine by Central Alaska Gold and Associates (now merged into Consolidated Nevada Goldfields Corp.) began in earnest in 1993. Based on the results of the exploration, Nevada Goldfields estimates there are "proven/probable" reserves of 122,549 tons of gold-bearing ore averaging 1.33 ounces of gold per ton that should yield 163,030 ounces of gold. The company also estimates an additional 39,160 tons of "possible" reserves yielding 37,600 ounces of gold.

To evaluate whether the mine would be profitable, the company estimated total capital costs for exploration and construction at about $15.5 million. Operating costs were projected at $169 per ounce of gold with gold valued at $385 per ounce. A gravity/flotation mill to be completed in 1995 is expected to process 150 tons of ore per day and produce 60,000 ounces of gold per year.

The mine's development has boosted the economy of local communities, which have supported its development. "We've probably spent in excess of $250,000 on freighting, groceries, lumber, wages and lodging in McGrath since we started work in 1989," said Chris Puchner, vice president of exploration.

With completion in 1995 of a large landing strip at the mine, money spent in the region on freighting, groceries, etc. will be reduced or eliminated. When asked how the mine can continue to contribute to the regional economy, Puchner points out that they have hired an average of five to six seasonal employees from the region each year and expect to hire that many full time once the mine is operating.

The mine's life, based on known potential reserves, is only a few years. However, the reserve estimates are based on less than 18 percent of the area thought to have potential for similar deposits. Also, the deepest drilling has been to 650 feet. With the total depth of potential gold resources extending to 5,000 feet, it is possible that Nixon Fork will be producing much longer. ◗

Ted Almsy of McGrath is part owner of the Nixon Fork claims. (Richard Montagna)

Hauling Freight

Story and photos by Richard Montagna

As in many remote mining operations in Alaska, material for the Nixon Fork Mine has to be freighted in using heavy equipment. The earlier mining operations barged their equipment from Seattle up the Kuskokwim River to Medfra, then overland on the trail to the mine. Equipment is now flown into McGrath, then freighted overland to Medfra and up to the mine. Usually two or more bulldozers will form a cat train, hauling the material on large sleds. Material is hauled to the mine in late winter or early spring when the river and tundra are frozen. The trail across the river is strengthened by drilling holes in the ice and bringing water to the surface to form an ice bridge safe enough to handle heavy loads.

CLOCKWISE FROM TOP RIGHT:
Neil Rosander of McGrath repairs the runner of a freight sled torn off by rough terrain while freighting supplies to the Nixon Fork Mine. The sled itself weighs about 20,000 pounds and was carrying 40,000 to 50,000 pounds of material.

A fully loaded sled awaits hauling to a mine site.

Chris Puchner, vice president for exploration for the Nixon Fork Mine, cinches down material for shipment to the mine.

L.E. Wyrick of McGrath heads out of town on his way to the Nixon Fork Mine. With warmer spring weather softening the trail in the daytime, he leaves at night to take advantage of colder temperatures and harder trail.

A Red Dog Becomes an Elephant

By Nick Jans

Editor's note: *Author of* Last Light Breaking *(1993), writer and teacher Nick Jans lives at Ambler in the Kobuk Valley.*

Alaska's remote northwest Arctic seems beyond time. Jagged peaks, ice-water rivers and tundra sweep to the horizon; caribou wander valleys where few humans have walked. Here, 90 miles north from Kotzebue in the DeLong Mountains, lies Red Dog, the state's most important hardrock mine. With estimated ore reserves of 85 million tons, and more discoveries expected, Red Dog may prove to be the largest lead and zinc mine in the world. Jointly operated by Cominco Ltd. of Canada and NANA, the mine, in full production since 1991, has become a major force on the global market. During its 50-year life span, it should yield more than half a million tons of high-grade zinc concentrate annually and contribute $5 billion to Alaska's economy.

Given its huge potential — it is what economic geologists call an "elephant" — Red Dog seems to have been destined for success. In fact, the deposit could easily still be lying dormant, untouched. As Mark Skok writes in the June 1991 issue of *Minerals Today*, "The Red Dog story tells of more than an incredibly large and high-grade ore body developed in one of the most harsh and inaccessible regions on earth....The mine project has brought together diverse groups on a scale rarely seen....From the beginning, cooperation — mixed with a healthy dose of optimism — has been the key to the timely development of Red Dog, despite what many industry experts suggested was an impossible undertaking."

The concentrate storage building, the largest building in Alaska, stands at the Red Dog port site near Kivalina on the Chukchi Sea coast. Zinc concentrate is trucked year-round along a 54-mile road from the mill site and stored here until the summer shipping season when the seas are free of ice. All concentrate is shipped to markets during the three-month shipping season. (Jeff Schultz, Cominco Alaska)

The deposit was spotted in 1968 by bush pilot Bob Baker, who reported it to Dr. Irving Tailleur of the U.S. Geological Survey. Baker prospected in his spare time and operated a small mining company that he called Red Dog, in honor of his pet dog, O'Malley, a reddish-colored mongrel. Two years after Baker died in a flying accident, the find was made public by Tailleur, who dubbed it Red Dog.

In 1975 Cominco geologists probed the area and quickly recognized the magnitude of the discovery. Although ownership of the land was clouded by the Alaska Native Claims Settlement Act, Cominco continued its exploration and eventually staked the deposit, in direct conflict with NANA, the region's Native corporation, who also claimed Red Dog. A protracted legal battle was resolved in favor of NANA, and the two corporate opponents settled their differences to become partners in 1982. Meanwhile, serious groundwork had continued: millions of dollars were invested in feasibility studies, development and construction options, a mile-long aircraft runway near the mine site, a search for investors. Geologists probed the vast sedimentary deposit, and more than half the test holes produced "spectacular" results.

Because the project was so huge and remote — 350 miles from the nearest highway — and the arctic environment so difficult, Cominco and NANA needed more than $400 million from investors to make the mine a reality. The state of Alaska allocated $175 million to fund the mine's transportation system, and a consortium of

LEFT: Red Dog uses the differential flotation process to produce zinc and lead concentrates at the mill. (Jeff Schultz, Cominco Alaska)

ABOVE: Trucks carry ore from the mine site to the mill site in the background. (Jeff Schultz, Cominco Alaska)

international banks helped with the rest.

The design called for an open-pit mine connected by conveyor to a nearby mill complex. After being run through a series of grinders and separators, the flourlike concentrate would be trucked 50 miles to a port site on the icebound Chukchi Sea, and stockpiled for barging during the brief ice-free season.

But before construction could begin, more than 90 permits were needed. Most of these concerned environmental protection, and not all were mere formalities. Clearance to build the port would, in fact, require an act of Congress to shift the boundaries of Cape Krusenstern National Monument, a sensitive and important archaeological area. Then, and only then, could the real work begin.

Although two decades had passed between discovery and development, once construction began in summer 1987, the pace was dizzying. From July to October, 11 barge loads of equipment and supplies were landed at the port site, bringing in more than a hundred bulldozers, loaders, scrapers and heavy trucks; 3 million gallons of diesel fuel; 3 million pounds of explosives for blasting; 4 million pounds in spare parts; $1 million worth of spare tires, numbers as huge as Red Dog itself. Enserch Alaska Construction spearheaded the road project, surging forward at the rate of 100 feet an hour, 20 hours a day in the bright arctic summer nights. Conditions were difficult, unstable tundra overlying permafrost, with nearly 500 stream crossings, what amounted to icy swampland. To save time and expense crossing such terrain, the road was engineered as a single lane with regular turnouts. First, a layer of geotextile fabric (special cloth spooled on huge rolls) was laid down to keep the road from sinking into permafrost-rich tundra; then a bed was built up with gravel quarried from pits along the route. More than 2 1/2 million cubic yards of fill and 2 million square yards of geotextile were needed, along with six miles of steel culvert and nine major bridges.

While racing the onset of winter and using machinery that could rip down a mountain, Enserch was compelled to follow strict environmental standards. Nothing outside the road and gravel pit corridors was to be disturbed; the land on each side would remain as it had for eons, marked only by game trails, wide as the sky, if forever redefined by the thin, dusty ribbon running through its heart.

By Thanksgiving there was a drivable road connecting the port and mine site. Enserch had come through 69 days ahead of schedule, at a cost of $1 million a mile.

Now Green Construction, the main contractor for the mine project, could go to work. The first crews began in January's bitter cold and darkness. By spring 1988 the operation was in full swing, with more than a dozen companies working long shifts in the growing warmth and light. A mile-long runway buzzed with activity; huge Hercules cargo planes ferried men and materials in a steady stream.

There were actually several main projects, all moving forward at breakneck speed. While heavy equipment crews stripped off overburden and prepared the actual deposit for development, others labored at a dam to contain the huge amount of waste water and runoff that the mine would create.

Meanwhile, the living-quarters complex grew at a seemingly exponential rate, assembled from 300 prefabricated modules built far to the south.

The mill itself, completed in 1990, is also modular; built in the Philippines, these eight units, some as large as 1,600 tons and eight stories high, were barged north to the port site, then moved up the haul road on gigantic crawlers, specialized vehicles of the type used by NASA to transport launching platforms. The port site, 50 miles to the west, was cast on an equally grand scale. The concentrate storage building, where the finely ground zinc is stockpiled, is the largest man-made structure north of the Arctic Circle, longer than four football fields and 11 stories high. Off to the side, dwarfed by comparison, stand three fuel tanks holding more than 3 million gallons each.

The grand opening was in 1990. Even though initial production began as planned, there were still major problems to be overcome. First, the world zinc price began a steady decline, due to lower than expected demand and oversupply on the world market. Also, the Red Dog ore, though rich — 17.1 percent zinc, 5.0 percent lead, plus traces of silver — proved more complex than first thought. To boost metal extraction to more profitable levels, Cominco installed $21 million of additional mining equipment. Another $11 million went into a ditching system to control runoff at the mine site, an environmental concern. Improved dust suppression systems have also been implemented, to protect workers from the potentially hazardous concentrate. "Safety and health standards are high," says Cominco's Ralph Eastman. "That's a continually monitored process."

In 1994, Red Dog yielded more than 2 million metric tons of high-grade lead, zinc and silver ore. The volume of zinc concentrate, 560,000 tons, was first in the world. Half of this fed Cominco's own smelter in Trail, British Columbia, and the remainder was divided equally between Europe and Asia.

Even so, the operation failed to show a profit. But an upswing in world prices and demand give both Cominco and NANA reason for optimism. "We foresee a good future as the price rises," says Eastman. "World inventories of zinc are dropping now, and should continue to do so."

While Red Dog has yet to prove the economic bonanza some thought it would be, its impact on the NANA region's economy has been huge. The operation employs roughly 350 people, many of them local residents; annual payroll to Inupiat workers exceeds $12 million, and NANA has received royalties up to $3 million. Says John Rense, general manager of NANA, "In an area that doesn't have many economic options, Red Dog's been really positive. Royalties have also provided a revenue base that allowed the Northwest Arctic Borough to form, and that's of vital importance to our region." [The Northwest Arctic Borough is the political unit that represents the NANA region.]

Perhaps the greatest measure of Red Dog's success is the cooperation and trust it's fostered between NANA and Cominco. Twenty years ago, wrangling in court, the two corporations seemed unlikely partners. Now they seem to have proven that similar joint ventures do indeed have potential. Charlie Curtis, president of NANA, says, "We feel fortunate to have a partner like Cominco. Other companies might have shut down and walked away when things got tough. Cominco hung in there."

Cominco's Ralph Eastman adds, "We're proud of the technology that Red Dog represents, but we're equally proud of the social contract we have with the people up there. We've shown you can work in that kind of environment, and make a go of it." ◗

BELOW: *This December view of the Red Dog Mine shows the mill, and at lower left the accommodations complex. (Ralph Eastman, Cominco Ltd.)*

LOWER RIGHT: *The concentrate is transferred from the port site on barges to waiting vessels anchored nearly four miles offshore. (Jeff Schultz, Cominco Alaska)*

The Rock Lady

By L.J. Campbell

Oro Holaday and Ivan Stewart made these portraits of each other in 1941, shortly before they married. (Courtesy of Oro Stewart)

As a young girl, Oro Stewart used to follow her mother around their cattle ranch in Idaho, where they'd look for interesting rocks.

Today, people follow Oro around looking for rocks. They follow this 78-year-old through the Amazon jungle, into the deserts of Australia, to India, Chile, Brazil, Turkey, Greece.

"It's fascinating. You never know when you'll find something new," Oro said recently, sitting in the living room of her Anchorage home.

Oro Stewart is a bona fide rock hound. She surrounds herself with rocks. Jade boulders fill the front yard outside. A tall geode of amethyst crystals sits like a sculpture on the floor beside her chair. Handcarved stone figures and polished rocks cover the coffee tables. There are rocks throughout Oro's rambling house, including a rock museum in the basement and a workroom equipped with a giant rock-cutting saw.

"Each rock is original, just the way nature made it. Nothing is duplicated," she explains. "And you never know what's inside until you cut it open."

Oro Stewart likes sharing the mystique of minerals with others. She's been active in the Chugach Gem and Mineral Society since 1962, serving as past president and field trip coordinator. She still occasionally leads rock-collecting trips in the state, in addition to her annual international excursions. She opens her rock museum at home to schoolchildren on request. And she's always willing to talk rocks when she's working at her store.

Stewart's Photo Shop. A landmark for 53 years in downtown Anchorage. It opened as a portrait studio during World War II, one of Anchorage's first. Today, it's a part camera store, part rock collector's paradise. Nikons, Leicas and Hasselblads share the narrow space with countless gems and minerals – dainty oval opals from Australia, lacy agate from Brazil, garnets and jade from Alaska. The front window display holds cameras and lens on one side, mineral exhibits on the other. A sign hanging under the awning outside reads, "Alaskan Jade Bulk by Lb."

The photo shop was Ivan Stewart's vision, something to keep his wartime bride busy. Ivan was an electrical engineer who whisked Oro from Washington state to Alaska in 1941. They had met through mutual friends at her boarding house in Pullman, where she was teaching college English. He wooed her in a whirlwind courtship layered with endearing practical jokes, then left that summer for Alaska. He worked construction at the military bases, planning to return with money to finish school. Instead, he wrote Oro and proposed she come to Alaska and marry him.

"I thought it wildly romantic," she recalls.

They set up housekeeping in Kodiak, where a naval base was going in. Ivan thought the soldiers needed pictures of themselves to send home, so he built a darkroom and opened a portrait studio. He had learned camera and darkroom skills one previous summer, and taught them to Oro. The shop was hers to run.

The next year, they moved to Anchorage so he could work at Elmendorf Air Force Base. They found a storefront downtown for their business and a house to live in, where Oro set out a huge flower and vegetable garden.

Oro took the portraits, developed the film and printed the images of soldiers, couples, children, families. She worked 12-hour days, often alone, while Ivan worked on base. It was her introduction to a place in Anchorage's history.

In 1949, she asked Ivan for a

pet, an Alaska pet. No dog for her. She wanted a reindeer. Oro found a herder in Nome willing to part with one. They named it Star because of the white patch on its forehead.

Today, Star is Anchorage's most photographed and best known pet. However, the current Star is Mrs. Stewart's fourth reindeer. The original Star died of old age at 23. The second Star lived only a year, stolen and butchered one night in the early 1980s by two men from Fairbanks, who were later caught and convicted; one served a year in jail. Star's mourners posted a sign on her pen, "We Love You Star." Star the third succumbed at age 2 to indigestion, thought to have been caused from eating plastic bags.

Then came today's Star, famous with Anchorage's children and tourists who visit her lot at the corner of 10th and L streets. Ivan tamed her with handfuls of lichens, apples and sugar, feeding her from a stool in her pen while winter winds howled, Oro recalls. Today, Star is protected by a burglar alarm that extends to the house, and a sign on her fence cautions, "Do not put plastic bags in Star's corral. They will kill her."

While Oro grew up learning about rocks from her mother, and later from geology courses in college, she didn't get involved in Alaska's mineral world until the early 1960s.

Ivan, among other things, was making 16mm films of Alaska to sell to movie companies, like Walt Disney. He decided to fly north of the Arctic Circle to near Kobuk and film the state's only jade mine on Dahl Creek. Ivan had done gold placer mining in Idaho as a teenager. When he found out the jade mine was for sale in 1962, he jumped to buy it. Oro blessed the deal and a new routine emerged. Ivan worked the claims in summer with a small crew, cutting and hauling out slabs of jade stone to work in the winter. He made table tops, floor and wall tiles, book ends, and sold stone in bulk.

Their new enterprise led them quite naturally into the folds of the newly organized Chugach Gem and Mineral Society. Under Ivan's presidency in 1970, the membership grew to about 800. They became respected in lapidary circles for their knowledge about Alaska minerals.

In 1964, they decided to move their store from the east end of Fourth Avenue to the west end, into a wooden building owned by Oscar Anderson. It was a fortunate move. Three months later, the Good Friday earthquake rocked Alaska and collapsed the building at their former location. They didn't escape completely in their new place. The street dropped out in front of the store, taking the sewer and plumbing. But they only lost about $12,000 of inventory. Ivan rebuilt what he could to get them back up and running.

In the meantime, they packed up some cameras and headed down the Seward Highway to photograph the damage. They reached Twentymile River bridge north of Portage, which had collapsed, and walked over it into the stricken town. The rising tide moved high over the subsided land, cutting off their return to the car. For several hours, they stood in the cold, talking with some others who also were stranded.

One of them was Mary Carey, a homesteader from Talkeetna who was working as a newspaper stringer. Oro had been one of the

BELOW: Ivan Stewart works with the first Star in 1960, training her to pull a sled down the streets of Anchorage. (Courtesy of Oro Stewart)

RIGHT: Ivan Stewart (bottom center) used to take 16mm movies of ice falling from the 400-foot-high ice wall during the Lake George breakup, at the head of the Knik River. Prior to the 1964 earthquake, the Knik Glacier formed an ice dam each winter, creating Lake George behind it. During spring thaws, the lake filled to overflowing, cutting a channel and bringing down huge ice falls. The lake's discharge in late summer usually brought flooding downriver. Ivan sold his films of this to Walt Disney Studios for use in the movie "White Wilderness." (Oro Stewart)

first people she'd met two years earlier, when she'd driven into Anchorage the first time. She'd stopped to pet Star and Oro had invited her into the house for coffee. Oro's rock collection sparked an exchange about their common interest in minerals, but Mary didn't tell her she would never go rock collecting again: Her husband had suffered a heart attack and died only months earlier during a rock collecting trip. After that initial encounter, Mary dealt with the Stewarts for business, to buy supplies for her home darkroom.

Mary was glad to see familiar faces amid the uncertainty of where the tide might end. Oro kept the conversation going, and as they talked she urged Mary to join the rock club. "She talked with such a calm voice, and I thought, 'darn it lady, if we live through this I'll join that club."

It was the start of a long, adventurous friendship. Mary, now the author of 11 books and a fiddlehead fern farmer, has gone with Oro on 25 foreign rock hunting trips.

"Oro is very soft spoken," says Mary. "You don't really realize she is leading. You feel free. She always tells you before you leave, what you expect to find, what we'll be looking for, why we're going to an area."

Most of the trips have been to places other people don't usually go, with experiences to match. They've slept in underground opal mines, visited tightly secured diamond mines, had their passports taken by communist border guards.

They've had countless other adventures in the state. They've driven the haul road, attempted to float the Koyukuk River in Oro's amphibian car, which she and Ivan had once floated down the Yukon, and attempted to kayak to Little Diomede in skinboats.

These days, Oro still keeps regular hours at the photo store, now with 11 employees.

She sold the jade mine after Ivan died in 1986, but little else has changed. She still has Star, as well as Ivan's big tabby cat, Amber. She still uses the 10-foot drag saw he designed to cut jade. She continues hosting the annual wild game feed, which they started serving free to the public years ago. People line up for blocks to file through the store for helpings of moose meat, seal, whale, muktuk, musk ox, buffalo, sheep, goat, salmon, char, trout.

She still gardens and fishes, even though arthritis pains her knees. She stays busy with rock club activities and planning her next overseas trip. At the store, she may be found at a desk in the back, hunched over paperwork. Or she may be planning a fishing trip with a friend, or talking rocks in one of the narrow aisles. Her knowledge about Alaska's rocks is encyclopedic.

"That's chrysanthemum rock," she says, pointing to a gray-green slice of stone patterned with darker green clumps like dabs from a paint brush. "It's feldspar crystals in limestone, but it looks like flowers. That's why I call it chrysanthemum rock. You get it down by Cordova, nine miles through the woods and swamps, by foot or swamp buggy, and carry it out." Next she picks up a fossilized mollusk from the Jurassic age.

A little boy at her elbow clutches an obsidian arrowhead and a chunk of polished jade that he's picked out from the rows of boxes of stones. It's hard to know whether he's most impressed by the jade that she mined, by her dinosaur-age fossils, or that she lives with Star the reindeer. He tells her Star is his friend. He tells her he wants to start collecting rocks and will come back here one day and get more. She makes him a deal: He can have the obsidian and jade if he'll bring Star an apple. "You want to encourage a budding rock hound, you know," she says, turning back into her store. ◆

BELOW: *Ivan and Oro Stewart were recognized for their many years of community service when they served as Fur Rondy Parade Marshals in 1985. (Courtesy of Oro Stewart)*

RIGHT: *In 1958, the Stewarts started an Anchorage tradition that continues today, a wild-game barbeque served free at Stewart's Photo Shop downtown during Fur Rendezvous. Here Oro Stewart greets some of the 1,200 people who filed through the store in 1994 for helpings of moose, seal, salmon, halibut and other types of game and fish. (Courtesy of Oro Stewart)*

In the Company of Rock Hounds

By L.J. Campbell

From an unassuming start in 1959, the Chugach Gem and Mineral Society grew to become one of the nation's largest rock clubs with a membership that once topped 800.

Today the society stays busy displaying its collection of minerals to groups around the state with weekly field trips in summer and twice monthly meetings year-round. The club puts on its biggest mineral show each February during Anchorage's Fur Rendezvous. Also each year, the club sponsors foreign rock collecting trips and funds scholarships of $200 to $1,000 to help Alaska college students pursue careers in geology and earth sciences.

For the Alaska centennial celebration in 1967, the society helped designate jade and gold as the state stone and state mineral.

Anchorage resident George Fennimore remembers when the society officially formed, meeting on Elmendorf Air Force Base at a time when getting there usually meant riding a bus. The club was started by geologists, federal employees and other rock hounds drawn together by their hobby. Of the 75 charter members, only a couple are still active today. Fennimore, age 86, is one of them.

He says that during World War II, years before the club's formal inception, miners and geologists used to get together and talk minerals at an assay office on Fourth Avenue. They'd show each other various rocks and trade secrets about what they'd found and where. At the time, Fennimore was prospecting claims on Sheep Mountain.

The Chugach Gem and Mineral Society rock show during Anchorage's Fur Rendezvous includes exhibits of Alaska jade from Dahl Creek, mined by Oro and Ivan Stewart. (Oro Stewart)

Once the society formed, club members raised money to buy rock cutting and polishing equipment and set up a lapidary shop on Elmendorf. Soon they were publishing a monthly newsletter called "Pebble Patter," which continues today. During the Alaska National Interest Lands hearings, when Congress was establishing new federal parks and preserves in Alaska, the society became politically active, lobbying legislators to keep lands open to mineral exploration and development.

Mostly, though, the society embodies a passion for exploring nature's elemental combinations. The club today has between 100 and 200 members, from jewelry makers, miners, geologists and mining engineers to people who just like rocks. Monthly speakers, field trips, special activities and events give members plenty of opportunities to swap stories, information and knowledge. Besides, rock collecting is fun.

"It's one of the best educations, something families can do," says Fennimore. "You're learning all the time about birds and animals and vegetation – lot of minerals you can tell just by the fungus on the rocks – and you're out getting fresh air."

The summer field trips reflect the members' special interests. Those particularly fond of fossils lead fossil collecting expeditions. Those more interested in gems take groups beachcombing for agates and jasper. Gold panning trips to the Kenai Peninsula and into the Talkeetna Mountains, garnet collecting trips to Wrangell, and expeditions to Kennecott copper country are perennial club adventures.

Along with the typical Alaska hiking provisions – water, rain gear, bug repellent and trail food – a rock collector's day pack also carries newspapers for wrapping specimens, a mosquito head net for boggy crossings, and, always, a chisel and rock hammer. Patience, too, goes with the territory.

"Some of the best specimens are found on the longer hikes," says Curvin Metzler, a computer science instructor and current society president. He recalls a recent club outing for fossils when, as usual, the group included novice collectors as well as old timers. "I was chipping away at a rock and a little girl came up, wanting to help, wanting to hit with the hammer. I ended up spending a little longer there, and as a result, found some snail fossils I wouldn't have otherwise." ◆

THE GALLOPING BUILDINGS OF FLAT

By Rolfe G. Buzzell

Editor's note: *Dr. Buzzell is an historian with the Office of History and Archaeology, Alaska Division of Parks and Outdoor Recreation. He is author of a forthcoming study about Flat's historic buildings and editor of a forthcoming volume of oral history interviews on Flat and Iditarod, both to be published by BLM.*

Mining in Alaska has impacted many towns, creating communities and then transforming them as the mining changed. Some of those communities, such as Iditarod, became ghost towns and a few, such as Innoko, have disappeared all together. Others have survived years of boom and bust cycles. Few communities in Alaska have been affected quite like Flat, which was the largest mining camp in the Iditarod District.

Flat, also known as Flat City, began as a

small mining camp at the confluence of Otter and Flat creeks in 1910. Located eight miles southeast of the town of Iditarod, Flat grew quickly because of its proximity to the richest placer ground in the district. As the town grew, tent frames and simple log buildings gave way to one- and two-story, wood-frame commercial buildings and residential dwellings. In the 1910s, Flat looked like many other frontier mining towns, with its classic street-scapes featuring false-front buildings, board sidewalks and muddy streets.

By the mid-1920s, Flat supplanted Iditarod as the supply center and the largest town in the district. It also became an important destination on the Iditarod Trail between Seward and Nome. Natural disasters and mining within the town's boundaries began changing the appearance of the community in the 1920s. A 1924 fire destroyed much of the town's business district. Periodic winter flooding, resulting from warm weather and glaciation on Otter Creek, prompted many business owners and residents to move their buildings to higher ground. The most significant physical impact on Flat and its buildings, however, came from the discovery of rich placer ground under the town itself in the mid-1920s. Flat was built on mining claims. Claim-holders and large-scale mining operators blocked efforts to incorporate the town. As a consequence, Flat was never platted. When dredge operators began mining portions of the town, they gave business owners and residents 10 day's notice to move their buildings. Dredging continued in the 1930s and later, and other sections of the town were forced to move. As dredging technology improved, portions of the town were dredged more than once in efforts to reach the gold deposits beneath the deep layers of overburden. Some buildings were moved two or more times as mining activity in the town continued through the 1960s.

Today, Flat is a mosaic of mostly abandoned buildings on tailings, resting on temporary foundations and widely dispersed among willow, alder and birch trees. Eighty-six buildings and structures in the town are associated with commercial, residential, public and mining activities. The buildings and structures represent a second or third generation mining camp, and most date from the 1910s to the 1930s. The frontier street-scapes of the 1910s and early 1920s have disappeared. Some buildings have collapsed or been torn down for the lumber. Those which remain exist because miners have used and maintained them as part of their operations. Flat has the largest concentration of gold rush era buildings and structures associated with the Iditarod National Historic Trail. Only one family resides year-round in the community, and Flat's historic buildings are linked with the uncertain future of mining in the area. ◆

FACING PAGE: *Flat, shown here about 1912 to 1914, was destined to grow into the largest community in the Iditarod Mining District. (Anchorage Museum, Photo no. B91.43.32; courtesy of Rolfe G. Buzzell)*

TOP RIGHT: *The Donnelley & Sheppard commercial complex fronts one of the main streets of Flat about 1930. To the right of the complex miners set thaw lines to prepare the ground for mining. (John Miscovich photo; courtesy of Rolfe G. Buzzell)*

RIGHT: *This aerial looking south over Flat in 1993 shows the helter-skelter pattern of building alignment that resulted from the Matheson's dredge's sweeps through the town site. (Rolfe G. Buzzell, Office of History and Archaeology, Alaska Department of Natural Resources)*

Terra *Incognito*

Story and photos by William Wakeland

Editor's note: *William Wakeland is a free-lance photographer-writer living in Anchorage and has been an Alaska resident since 1946.*

Robert Marshall described them as "...a series of sensational needlelike peaks extending for six or eight miles in a horseshoe around a gushing creek..." in *Arctic Wilderness* (1956). However, his interest had been aroused before he'd even heard of the Arrigetch Peaks. After exploring the central Brooks Range to the east, he offered this observation: "By July, 1931, there yet remained in the Arctic Koyukuk drainage a 'terra incognito' forty five by a hundred square miles in extent...almost as unknown as the geography of the moon." So he and a friend set out from Wiseman to remedy that situation, with a boat, motor, pack dogs and 50 days of supplies.

This part of Gates of the Arctic National Park lies west of the Alatna River, north of Walker Lake and east of the Noatak River headwaters. The only previously known exploration by other than Natives was 20 years earlier by geologist Philip Smith, for whom a mountain range to the east is named. Neither he nor Marshall seem to have more than glimpsed the Arrigetch Peaks, which are not readily visible from the Alatna River, the principal avenue of travel.

Among the early explorers of this area were Bud and Connie Helmericks, who had canoed part way up the Alatna in 1945. In *The Flight of the Arctic Tern* (1952), Connie describes "finding" Takahula Lake from the air after they had returned with an airplane. The lake wasn't on their maps but was only a couple hundred yards from the Alatna, and Connie and Bud built a cabin there in the early 1950s. Takahula marks the southeasterly corner of the Arrigetch region and was the pick-up point for my last hike. There is still a cabin on the lake.

One of several Matterhorn-shaped pinnacles in the area, this peak is viewed over a shed caribou antler on upper Arrigetch Creek.

Backcountry hikers love this area both for the scenery and for the lack of people, thanks to the time, expense and effort to get there. The peaks are best reached by hiking from water landings on or close to the Alatna River, after a flight by bush plane from Bettles, an hour away to the southeast, which in turn is reached by air from Fairbanks. In addition to the expense and effort to get there, further protection for the peaks is provided by the three B's: bugs, bears and brush, at least in the summer at the lower elevations. Takahula and Circle lakes are only about 900 feet elevation.

My last hike there, with a group from the Mountaineering Club of Alaska, started at Circle Lake, a former meander of the Alatna River. Because it was late in the day, we camped on the brushy, buggy sidehill, only to be roused early in the morning by yells and cursing. There was a black bear in camp. We ran him off, but decided to get out of there before he returned. This was an "experienced" bear — thanks to careless hikers in the past — lured to camp by the plane that dropped us off.

The Alatna Valley in this area is broad, largely spruce-timbered and

Chris Zafren gazes into a cottongrass-bordered pool on the main fork of Arrigetch Creek above the forks at about 3,000-foot elevation.

wet in many places. A trail of sorts meanders north from Circle Lake toward the mouth of Arrigetch Creek, but we angled up the hillside to find our way on firmer ground to a ridge overlooking the creek, where there's a trail shared with animals that parallels the creek. After a miserable start sweating our way up the hillside in mosquito nets and rain gear, things got better as we gained altitude and caught our first glimpse of the peaks. From a campsite with a view of the peaks, the trail extends another five miles southwesterly up the creek to a major fork. Near this fork, at the base of Twin Peaks 3805 (our name, plus the elevation shown on the U.S. Geological Survey map), we made our base camp for several days because we had good water, easy hiking, great scenery, and at 2,500 feet elevation, fewer bugs.

From this camp we could hike up three of the Arrigetch forks into the alpine valleys. At this point also turquoise-colored Arrigetch Creek has gentled out and is easily crossed most of the time in summer.

The most visible features of the Arrigetch region, such as the peaks, watersheds and glaciers, have been named by climbers in their reports in the *American Alpine Journal*. Some of the names are rather common, but the latest USGS inch-to-a-mile series maps name only Arrigetch Creek, plus the Alatna River and Takahula Lake. Names such as Wichmann Tower, Pyramid, Battleship, Citadel and Maiden, found in *Journal* accounts, do not appear on maps, and valleys are numbered. So our names will have to do here, even though readers know the names are not official. Some, like The Maidens, are known widely. But what we call Battleship, Twin Peaks or Matterhorn may not be the same ones so named by alpine climbers, and there are several peaks fitting each of these names.

Marshall found that Natives had quite recently named the rivers and streams. In 1931 he had encountered two Eskimos along the Alatna who had not seen another human for six months. From them he learned some of the names now found on maps, including Arrigetch. These names were given by families of a current or immediate past generation. Names like *Pingaluk*, a tributary from the east, which means "a little no good" and refers to the fishing. Or *Unakserak*, the tributary opposite Arrigetch Creek, which means "a place where one goes to get wood for snowshoes."

From our base camp at the forks, we could now explore the valleys with day packs. The longest, and perhaps most picturesque, fork is the main branch. Faint trails and shed horns and antlers are evidence of sheep, caribou and moose. At this time of the year we see only a few sheep. A mile above camp, the valley widens and the creek meanders in and out of pools bordered by saxifrage, cottongrass and dwarf fireweed, pastoral scenes in sharp contrast to the near vertical spires above. A

Hikers cross upper Arrigetch Creek in a slow-moving spot at low water, but the crossing is still treacherous because of deep holes hidden under the glacial water.

High up on Arrigetch Creek and overlooking a peak dubbed The Ship, hikers rest on a glacial field that tops out above 7,100 feet.

couple more miles and the most interesting shapes loom before us. Our favorite we call The Ship; on the map it is "5565." A mile to the west is "6685," a sharp cone we had dubbed The Matterhorn. Whether the snowfields now visible hide true glaciers is a mystery, but I'm sure some of them do.

After several more day hikes up other tributaries, we moved camp to the east a few miles, staying above timberline. The new site is a delight, overlooking a small lake high in a brush-free valley with several new peaks to gaze at. That evening a blond grizzly approached the lake from one end, gave us a rather disdainful glance, waded in and swam to the other end, scratched its rear on a rock, shook off the water, glanced at us again and wandered off.

We left that valley reluctantly, after some hikes, berry-picking and lying around in the sun. A long descent, largely on caribou trails, down into timber brought us to an unnamed creek. We called it Hot Springs Creek, simply because the old one-inch-per-four-miles map shows a hot springs on the map. We found it, even though it's not on the current maps, and it's really hot. The stream in this new valley is as big as Arrigetch Creek, a real torrent after a rain as we soon learned, yet it doesn't even have a name on the latest maps.

After a night and day of rain at this base camp no. 3, we were relieved to find the creek down considerably next morning. We had to cross the creek to reach our pick-up point on Takahula Lake, and succeeded less than a mile below camp. Good thing we did, too, because we had another couple days of rain for a long, brushy, wet drag to our destination.

An account in the 1963-1964 *Alpine Journal* chronicles another expedition using Takahula Lake as an access point. That party had a bush plane drop their primary food supplies and climbing gear near, or on the way to, a formation they called the Citadel. Then the party landed at Takahula Lake and proceeded to hike in. According to the report in *Alpine Journal*, "...having struggled with loaded Keltys [a brand name for a backpack] for 50 hours through thick alders, over sharp, loose scree, through hordes of mosquitoes, and across glacial meltwater streams, [we] finally reached the point over which they had released the drop." They couldn't find any of the drop and had to retreat to Takahula Lake where they eventually flagged a passing plane.

By comparison, our trek to the lake was a lark. We got there just as the rain let up, and found a wilderness paradise, even the bugs had nearly quit. The sun came out and we draped wet clothing all over the place before jumping into the lake for a swim. That evening we watched fish jump as ducks paddled close, a moose waded out and loons tuned up.

Takahula Peak, nearly 3,500 feet, stands guard over the lake, reminding us we are still close to the Arrigetch Peaks. We bid them farewell from aloft in our chartered Beaver next day on our flight back to Bettles. ■

Tenacious Telida

Story and photos by Richard Montagna

"Heavy snowfall isolates village, Telida residents have little food, gas"
–*Anchorage Daily News,* Feb. 1, 1993

"Subzero weather stymies supply flights to Native village"
–*Anchorage Daily News,* Feb. 3, 1993

For many Alaskans, these newspaper stories were the first they had heard of Telida, a small Athabaskan village on the Swift Fork of the Kuskokwim River. The village had received statewide attention when heavy snowfall and deep cold isolated it from the outside world in February 1993. Delivery of needed gasoline purchased in the fall had been delayed. Heavy snowfall and broken snow removal equipment had left the runway closed to all but ski planes. Without gasoline, the villagers were unable to pack the runway or break trail to Nikolai 50 miles away. Food supplies were dwindling fast and minus-40- to minus-50-degree weather was preventing ski planes from reaching the village.

The weather eventually warmed up enough for fuel to be delivered and residents headed to Nikolai and on to McGrath for supplies. Later that year, the village received a legislative appropriation to fix the bulldozer needed to keep the runway open. But even with this crisis solved, the village continues to face a tenuous future.

Telida was first visited by non-Natives at the turn of the century. In 1899, a government expedition led by Lt. Joseph Herron was searching for an overland route from Cook Inlet to the Yukon River. Lost and weary, they wandered up the North Fork of the Kuskokwim where they were rescued by Chief Sesui and his band from Telida. Expedition members spent a couple of months recuperating in the band's village.

Changes in the river's course forced periodic relocation of the village. The site where the Herron party rested was soon abandoned, and the band moved to what is now known as Old Telida. Then, in 1916, some residents moved four to five miles downstream to New Telida, site of the present-day village. Old Telida was finally abandoned in 1935 when the village flooded.

Although hundreds of people passed through Telida when it was a stopover on the McGrath-Nenana Trail in the early 1920s, the resident population has always been relatively small. Lt. Herron noted only 17 residents from four families lived here in 1898; seven residents were reported in 1935. In 1949 there was one resident family, the Carl Sesui family. In 1960, three families lived here; the 1980 census counted 33. The current population of Telida has now

Telida's only source of income comes from selling power it produces to residents. Individuals rely on subsistence supplemented by trapping and some fishing to generate cash. But since the mail flight has been cancelled, chartering a flight has dramatically boosted the cost of groceries and supplies.

LEFT: Finding enough students to keep its school open is one of the most immediate concerns for Telida residents. The state is raising its minimum enrollment standards and the community's youngsters may be placed in a correspondence program if enrollment doesn't increase or some other arrangement can't be worked out with the Iditarod Area School District.

LOWER LEFT: Alaska has many remote communities, but few are so isolated that their very existence is threatened. Telida, an Athabaskan outpost on the Swift Fork of the Kuskokwim River, faces just such a challenge.

declined to one permanent family: Steve and Irene Nikolai and their five children. Steve Eluska sometimes resides here; other times he stays with his wife, Olga, who teaches outside the village. Other frequent residents are Steve's parents Heldina and Deaphon Eluska of McKinley Fork, a creek in the Telida area. The only other family at the village, five members strong, is that of the schoolteacher.

The village itself only makes money from producing the power that it sells. Since government-funded services to a community are usually based on population size, the decrease in the population of the village during the last 10 years has brought reductions in services most people take for granted and an uncertain future for the village.

One of the services that has been eliminated is the delivery of mail to the village. Weekly mail service from McGrath had been initiated in the mid-1970s. Now, although Tanana Air flies past Telida on their twice-weekly flight between McGrath and Fairbanks, the village is too small for the Postal Service to pay for a scheduled mail stop. Previously, residents were able to order supplies from the store in McGrath and have them shipped in on the mail plane. Now they must charter a flight from McGrath at $290 to $320 every few weeks to get groceries and mail. Without a scheduled mail run, villagers must also depend on charters for air transportation out of the village. Seat fare on a mail run would cost $70 to McGrath in spring 1995, instead of the current charter rate.

In the past, one of the villagers was employed by the Tanana Chiefs Conference as a primary health aide trained in village health surveillance and preventive health care. The village council built a clinic that Indian Health Services maintained for providing medical services. When the health aide move away in the late 1980s, the clinic was closed and health care to the village discontinued.

Schooling in Telida takes place in a one-room schoolhouse, with all students from kindergarten through high school studying in the same room with one teacher. While this type of learning environment can have its drawbacks, it can also present unique opportunities for the students. Peggy Dargon, an infant-learning teacher with Four Rivers Counseling of McGrath, notes her impressions from visits to the school in Telida. "I was really impressed with the way the kids knew their environment. They knew a lot about the land, the animals and the river. The curriculum seemed to be designed for the kids. They were free to learn by exploring. The teacher's kids from outside the village complemented the kids from the village. They learned from each other."

When asked about curriculum and how it might differ from that in a larger community, first-year teacher Shelly Carpenter noted that the district had created work units based on seasonal activities the villagers do, their local environment, local animals etc....that the teacher had the option of using. Shelly, who is originally from McGrath, talked about living and teaching in Telida. "The greatest joy of teaching here is really getting to know the students and their family. It's like a family atmosphere. Everyone is genuinely concerned about everyone else. The students are able to utilize their own environment and culture in their learning activities. Our own two children and the others are able to learn at their own pace. They also have a real feel of ownership of the school. They feel it is their own school; it's a place they're very comfortable in."

Whether the students in Telida will continue to have a school of their own is questionable. Under changes in the administrative policy of the state Department of Education, the Iditarod Area School District, of which Telida is a part, may have to face the challenge of finding funds to keep the school open and provide its eight students with quality education. For the 1995-1996 school year, the minimum number of students required to keep a school funded by the state will increase from eight to nine and to 10 the following year. If enrollment is below these minimums, the IASD Board will have to decide whether to keep the school operating out of monies received for the other schools in the district or to close the school and enroll the children as McGrath students for funding and teach them through a correspondence program.

The children seem to make the most of where they live, according to Shelly. "Unless the weather is bad, they usually play outside until it's dark, and the older kids play with the young ones, which is nice."

When asked about the future of his village and if he thought his children might want to live somewhere else when they get older, Steve Nikolai answered: It's home to them, they don't want to move to another place." In a later conversation, his children echo his sentiments: "It's quiet here and there's lots to do outside," said Jim, age 13. Like his father, who lists trapping as his main source of income, Jim enjoys the lifestyle available here. "I like going out in the woods and I like to trap, but I only got five (marten) this year."

Nine-year-old Steven enjoys going to school where he can learn to draw and play with computers. "I like art the best. In the summer I like fishing the most," he says, and quickly names the types of salmon and other fish he catches. When asked what he thinks he will do when he is older, he too talks about trapping, fishing and boating.

Steve Nikolai hopes his family will continue to call Telida home. Asked about changes he would like to see, he replies, "Getting a mail run once a week would be nice." As for solutions to the village's problems, his response is simple, "We need a baby boom to get more people here." ■

The Russian Orthodox chapel of St. Basil the Great, originally built in Old Telida in 1918, was moved to the new village site log by log.

Bibliography

Associated Press. "Greens Creek gets new life." *Anchorage Daily News*, May 18, 1995.

Bales, Carole A. and Sonja K. Runberg, eds. "Understanding Our Planet Through Chemistry." On-line document compiled by Joseph E. Taggarts, Jr., U.S. Geological Survey, 1995.

Barkeley, James N. "American Gold: Facts & Figures." *The Alaska Miner.* Anchorage: Alaska Miners Association, April 1991.

Brooks, Alfred H. *Blazing Alaska's Trails*. Fairbanks: University of Alaska Press, 1953.

Bundtzen, T.K., R.C. Swainbank, A.H. Clough, M.W. Henning and E.W. Hansen. *Alaska's Mineral Industry 1993*. Anchorage: Division of Geological & Geophysical Surveys, Special Report 48, 1994.

Consolidated Nevada Goldfields Corp. Quarterly Reports. Denver, Colo.: unpublished, 1995.

"Fairbanks, The Golden Heart City." *The Alaska Miner*, Vol. 21, No. 6, June 1993.

Garnett, Richard. "Development of an Underwater Vehicle for the Offshore Placer Gold Deposits of Alaska." *The Alaska Miner*, Anchorage: Alaska Miners Association, July and August, 1992.

Herried, Gordon. *Geology and Geochemistry of the Nixon Fork Area, Medfra Quadrangle, Alaska*. Juneau: Alaska Division of Mines and Minerals Geologic Report No. 22, June 1966.

Higgs, Andrew S., Robert A. Sattler. *History of Mining on Upper Fish Creek*. Fairbanks: Northern Land Use Research, Inc., 1994.

Jasper, M.W. *Report on the Mespelt Mine of Strandberg Mines, Inc. Nixon Fork District, Medfra Quadrangle Alaska*. Report PE 65-1. Juneau: Alaska Division of Mines and Minerals, 1961.

LeFebvre, Dick. "Focus On...Fort Knox." *Alaska's Land*. Juneau: Department of Natural Resources, 1992.

Morgan, Albert Weldon, edited by Rolfe G. Buzzell. *Memories of Old Sunrise*. Anchorage: Cook Inlet Historical Society, 1994.

Morgan, William. *Report on Nixon Fork Mining District, Medfra Quad*. Anchorage: Alaska Duval Corp., unpublished report, 1983.

Project Description for the Fort Knox Mine. Fairbanks: Fairbanks Gold Mining, Inc., 1992.

Rennick, Penny, ed. *Alaska's Oil/Gas & Minerals Industry*. Anchorage: The Alaska Geographic Society, 1982.

—. *Juneau*. Anchorage: Alaska Geographic Society, 1990.

U.S. Geological Survey. *The Petroleum Fields of the Pacific Coast of Alaska*, Bulletin 250. Washington, D.C.: U.S. Geological Survey, 1905.

Wernecke, Livingstone. *Report on Nixon Mines, Mount McKinley Recording District, McGrath, Alaska*. Treadwell, Alaska: Alaska Treadwell Gold Mining Co., unpublished report, Dec. 28, 1920.

Index

Photographers

ALASKA GEOGRAPHIC. Back Issues

The North Slope, Vol. 1, No. 1. Out of print.
One Man's Wilderness, Vol. 1, No. 2. Out of print.
Admiralty...Island in Contention, Vol. 1, No. 3. $7.50.
Fisheries of the North Pacific, Vol. 1, No. 4. Out of print.
Alaska-Yukon Wild Flowers, Vol. 2, No. 1. Out of print.
Richard Harrington's Yukon, Vol. 2, No. 2. Out of print.
Prince William Sound, Vol. 2, No. 3. Out of print.
Yakutat: The Turbulent Crescent, Vol. 2, No. 4. Out of print.
Glacier Bay: Old Ice, New Land, Vol. 3, No. 1. Out of print.
The Land: Eye of the Storm, Vol. 3, No. 2. Out of print.
Richard Harrington's Antarctic, Vol. 3, No. 3. $17.95.
The Silver Years, Vol. 3, No. 4. $17.95.
Alaska's Volcanoes, Vol. 4, No. 1. Out of print.
The Brooks Range, Vol. 4, No. 2. Out of print.
Kodiak: Island of Change, Vol. 4, No. 3. Out of print.
Wilderness Proposals, Vol. 4, No. 4. Out of print.
Cook Inlet Country, Vol. 5, No. 1. Out of print.
Southeast: Alaska's Panhandle, Vol. 5, No. 2. Out of print.
Bristol Bay Basin, Vol. 5, No. 3. Out of print.
Alaska Whales and Whaling, Vol. 5, No. 4. $19.95.
Yukon-Kuskokwim Delta, Vol. 6, No. 1. Out of print.
Aurora Borealis, Vol. 6, No. 2. $19.95.
Alaska's Native People, Vol. 6, No. 3. $24.95.
The Stikine River, Vol. 6, No. 4. $17.95.
Alaska's Great Interior, Vol. 7, No. 1. $17.95.
Photographic Geography of Alaska, Vol. 7, No. 2. Out of print.
The Aleutians, Vol. 7, No. 3. Out of print.
Klondike Lost, Vol. 7, No. 4. Out of print.
Wrangell-Saint Elias, Vol. 8, No. 1. Out of print.
Alaska Mammals, Vol. 8, No. 2. Out of print.
The Kotzebue Basin, Vol. 8, No. 3. Out of print.
Alaska National Interest Lands, Vol. 8, No. 4. $17.95.
Alaska's Glaciers, Vol. 9, No. 1. Revised 1993. $19.95.
Sitka and Its Ocean/Island World, Vol. 9, No. 2. Out of print.
Islands of the Seals: The Pribilofs, Vol. 9, No. 3. $17.95.
Alaska's Oil/Gas & Minerals Industry, Vol. 9, No. 4. $17.95.
Adventure Roads North, Vol. 10, No. 1. $17.95.
Anchorage and the Cook Inlet Basin, Vol. 10, No. 2. $17.95.
Alaska's Salmon Fisheries, Vol. 10, No. 3. $17.95.
Up the Koyukuk, Vol. 10, No. 4. $17.95.
Nome: City of the Golden Beaches, Vol. 11, No. 1. $17.95.
Alaska's Farms and Gardens, Vol. 11, No. 2. $17.95.
Chilkat River Valley, Vol. 11, No. 3. $17.95.
Alaska Steam, Vol. 11, No. 4. $17.95.
Northwest Territories, Vol. 12, No. 1. $17.95.
Alaska's Forest Resources, Vol. 12, No. 2. $17.95.
Alaska Native Arts and Crafts, Vol. 12, No. 3. $22.95.
Our Arctic Year, Vol. 12, No. 4. $17.95.
Where Mountains Meet the Sea, Vol. 13, No. 1. $17.95.
Backcountry Alaska, Vol. 13, No. 2. $17.95.
British Columbia's Coast, Vol. 13, No. 3. $17.95.
Lake Clark/Lake Iliamna, Vol. 13, No. 4. Out of print.
Dogs of the North, Vol. 14, No. 1. $17.95.
South/Southeast Alaska, Vol. 14, No. 2. Out of print.
Alaska's Seward Peninsula, Vol. 14, No. 3. $17.95.
The Upper Yukon Basin, Vol. 14, No. 4. $17.95.
Glacier Bay: Icy Wilderness, Vol. 15, No. 1. Out of print.
Dawson City, Vol. 15, No. 2. $17.95.
Denali, Vol. 15, No. 3. $19.95.
The Kuskokwim River, Vol. 15, No. 4. $17.95.
Katmai Country, Vol. 16, No. 1. $17.95.
North Slope Now, Vol. 16, No. 2. $17.95.
The Tanana Basin, Vol. 16, No. 3. $17.95.
The Copper Trail, Vol. 16, No. 4. $17.95.
The Nushagak Basin, Vol. 17, No. 1. $17.95.
Juneau, Vol. 17, No. 2. Out of print.
The Middle Yukon River, Vol. 17, No. 3. $17.95.
The Lower Yukon River, Vol. 17, No. 4. $17.95.
Alaska's Weather, Vol. 18, No. 1. $17.95.
Alaska's Volcanoes, Vol. 18, No. 2. $17.95.
Admiralty Island: Fortress of Bears, Vol. 18, No. 3. $17.95.
Unalaska/Dutch Harbor, Vol. 18, No. 4. $17.95.
Skagway: A Legacy of Gold, Vol. 19, No. 1. $18.95.
ALASKA: The Great Land, Vol. 19, No. 2. $18.95.
Kodiak, Vol. 19, No. 3. $18.95.
Alaska's Railroads, Vol. 19, No. 4. $18.95.
Prince William Sound, Vol. 20, No. 1. $18.95.
Southeast Alaska, Vol. 20, No. 2. $19.95.
Arctic National Wildlife Refuge, Vol. 20, No. 3. $18.95.
Alaska's Bears, Vol. 20, No. 4. $18.95.
The Alaska Peninsula, Vol. 21, No. 1. $19.95.
The Kenai Peninsula, Vol. 21, No. 2. $19.95.
People of Alaska, Vol. 21, No. 3. $19.95.
Prehistoric Alaska, Vol. 21, No. 4. $19.95.
Fairbanks, Vol. 22, No. 1. $19.95.
The Aleutian Islands, Vol. 22, No. 2. $19.95.
Rich Earth: Alaska's Mineral Industry, Vol. 22, No. 3. $19.95.

ALL PRICES SUBJECT TO CHANGE

Your $39 membership in The Alaska Geographic Society includes four subsequent issues of *ALASKA GEOGRAPHIC®*, the Society's award-winning quarterly. Please add $10 per year for non-U.S. memberships.

Additional membership information and free catalog are available on request. Single *ALASKA GEOGRAPHIC®* back issues are also available. For back issues add $2 postage/ handling per copy for Book Rate; $4 each for Priority Mail. Inquire for non-U.S. postage rates. To order back issues send check or money order (U.S. funds, please) or credit card information (including expiration date and daytime phone number) and titles desired to:

ALASKA GEOGRAPHIC.

P.O. Box 93370 • Anchorage, AK 99509-3370
Phone (907) 562-0164; fax (907) 562-0479

NEXT ISSUE: *World War II in Alaska*, Vol. 22, No. 4. Fifty years ago it ended, but the memories continue, of the war that mobilized the nation's military to defend its northern frontier. Southeast, Southcentral, the Interior, the Aleutians, and the Bering, Chukchi and Arctic coasts... no part of Alaska was untouched. The Alaska Highway, improved communications, more airstrips and seaports, all are legacies of the military build-up that shaped modern Alaska. This issue commemorates Alaska at war. To members 1995, with index. Price $19.95.

ALASKA GEOGRAPHIC® 1995 Holiday Shopping Guide

ALASKA TREATS from ALASKA SAUSAGE & SEAFOOD!

Send a taste of Alaska to friends and family this year with these delectable treats from Alaska Sausage and Seafood. Items are drop shipped from Anchorage by Federal Express.

Reindeer Sausage Ring: This popular 12-ounce sausage ring is made from pork, beef and reindeer meat, then smoked in a natural casing with hearty spices.

Seafood Snack Pack: This twin pack contains seasoned, smoked and fully cooked Alaska halibut and salmon (6 ounces of each).

Prices (plus shipping/handling, below):
Reindeer sausage, $5.95 [members: $4.76];
Seafood pack, $15.95 [members: $12.76].
S/H (sent overnight via Federal Express):
Add $10 for the first item, $2 for each additional item sent to same address.
Sorry, no foreign orders for these items.

ALASKA GEOGRAPHIC® GUIDES

Once called "Paris of the Pacific," today Sitka is one of Southeast Alaska's most popular destinations. Nestled between ocean and rain forest on the coast of a mountainous island, this busy fishing port and burgeoning cultural center boasts unsurpassed scenery and boundless history.

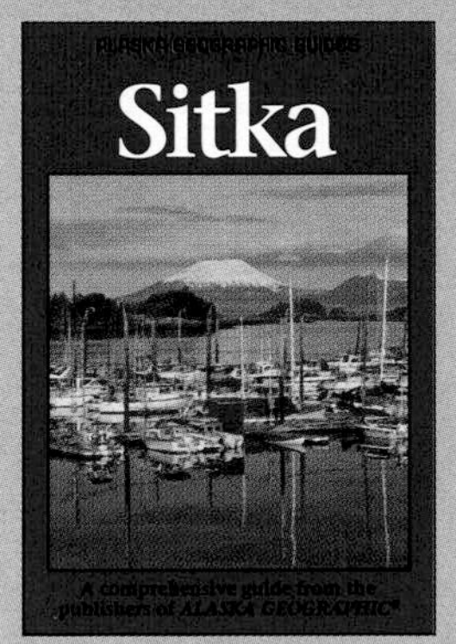

In recognition of this, Alaska Geographic proudly presents ***Sitka***.

This new, comprehensive guide offers a close look at Sitka, its natural environment, history as a Tlingit stronghold and capital of Russian America and provides suggestions of things to see and do, restaurants to try, even a list of kids' picks for families. Gorgeous color photos and maps complement the text.

Don't miss ***Sitka***, the first in a new series of *ALASKA GEOGRAPHIC® GUIDES*, available 12/95. 160 pages with index.

Price: $15.95 [members: $12.76] plus S/H (per book: U.S.–$2 book rate, $4 Priority Mail; foreign–$4 per book surface mail)

ALASKA ULU KNIVES from The Ulu Factory

Alaska Natives have used ulu (pronounced oo-loo) knives for centuries. These high-quality stainless steel knives are made today by The Ulu Factory in Anchorage. The ulu's unique design makes it easy to hold and use.

(Birch shown)

Inupiat Style Ulus (left) are 8" wide, with large, easy-to-grip wooden handles and an open blade, and are available in walnut or birch.

The 6" **Ulu Steak Knife Set** (at right) includes four individual walnut-handled ulus with a single four-slot stand.

All ulus come with matching wooden stands and a history/instruction booklet.

Prices (plus S/H charges, below):
Inupiat Style Ulu, $19.95 [members: $15.96], (choose walnut or birch handle);
Ulu Steak Knife Set, $59.95 [members $47.96].
Shipping/handling: Add $3 per item.
Sorry, no foreign orders for these items.

PREHISTORIC ALASKA POSTERS

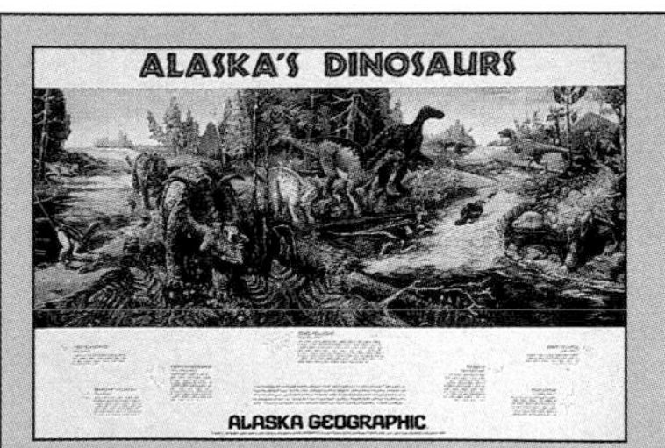

Dinosaurs and other animals that roamed the North in prehistoric times come to life on these colorful posters, made from Tom Stewart's original paintings. The posters also make perfect companions to *ALASKA GEOGRAPHIC®* Vol. 21, No. 4, *Prehistoric Alaska*. Choose **Alaska's Dinosaurs**, **Alaska's Ice Age Mammals** or the **Poster Set** (one of each) ; mailed in a tube.

Price: $12.50 [members: $10.00] each; $22.50 [members: $18.00] per set, plus S/H: U.S.–$4 for every 2 posters ordered; foreign–$9 per 2 posters (air mail)

— ORDER FORM —

ADDRESS #1

YOUR NAME ____________________

ADDRESS ____________________

CITY/STATE/COUNTRY ____________________ ZIP/POSTAL CODE ________

MESSAGE FOR GIFT CARD ____________________

ADDRESS #2

GIFT RECIPIENT'S NAME ____________________

ADDRESS ____________________

CITY/STATE/COUNTRY ____________________ ZIP/POSTAL CODE ________

MESSAGE FOR GIFT CARD ____________________

Please send these items to the addresses indicated (write address number from above in the "ship to" column below; attach a separate sheet for more gift recipient addresses):

QTY	ITEM	SHIP TO #	PRICE (incl. S/H)
	Reindeer Sausage Ring		
	Seafood Snack Pack		
	SITKA guide		
	Inupiat Style Ulu – birch handle		
	Inupiat Style Ulu – walnut handle		
	Ulu Steak Knife Set		
	Alaska's Dinosaurs poster		
	Alaska's Ice Age Mammals poster		
	Prehistoric Alaska Posters (set of 2)		
	The *MILEPOST*®		
	Denali, Vol. 15, No. 3		
	Wild Critters		
	Dance on a Sealskin		
	ALASKA GEOGRAPHIC® subscriptions ($39/year U.S.; $49/year to foreign addresses)		
		ORDER TOTAL:	$

— PAYMENT OPTIONS —

☐ Check/money order enclosed (U.S. funds only)

☐ Please charge my ☐ VISA or ☐ Mastercard:

Card # ____________________ Exp. date ________

Signature: ____________________

Daytime phone: ____________________

Clip or copy this form and return with your payment or credit card information to:

ALASKA GEOGRAPHIC SOCIETY
P.O. Box 93370–XA95 • Anchorage, AK 99509

The MILEPOST®

1995-96 Edition

"The bible of North Country travel since 1949"

The MILEPOST®, the most complete, comprehensive guide to travel in the North, makes a perfect gift for anyone planning or even contemplating a trip to Alaska.

This indispensable traveling companion provides mile-by-mile logs and detailed maps of all northern highways, including the Alaska Highway, plus information on ferries and cruise ships, accommodations, food, camping and what to see and do along the roads and in communities throughout the North. 722 pages, with 21" x 31" fold-out map and index.

Price: $19.95 [members, $15.96], plus S/H (per book: U.S.–$4 book rate; $6 Priority Mail; foreign–$9 surface; $19 air mail)

BACK IN PRINT!!

Denali,

Vol. 15, No. 3

Crown of the Alaska Range, 20,320-foot Mount McKinley has fascinated man for centuries. This stunning issue of *ALASKA GEOGRAPHIC*® provides an in-depth look at the mountain, its lofty neighbors and the surrounding parklands and wilderness areas.

Included are chapters on the history of the mountain; climbing Denali; mining in the Kantishna Hills; building the railroad through this rugged land; a guide to the Denali National Park road; and information on accommodations, transportation and activities. 96 pages, full-color throughout, with index and fold-out map.

Price: $19.95 [members, $15.96], plus S/H (per book: U.S.–$2 book rate; $4 Priority Mail; foreign–$4 surface; $12 air mail)

KIDS • KIDS • KIDS • KIDS • KIDS • KIDS • KIDS • KIDS • KIDS • KIDS

WILD CRITTERS: Tim Jones' whimsical verse and Tom Walker's photographs of Alaska's animals at work and play will amuse children of all ages. 48-pages, hardbound, 11 1/2" x 8 1/2". (Ages 3+)

Price: $15.95 [members, $12.76], plus S/H (per book: U.S.–$2 book rate; $4 Priority Mail; foreign–$4 surface; $12 air mail)

DANCE ON A SEALSKIN: With the help of Teri Sloat's colorful illustrations, author Barbara Winslow tells the heartwarming story of an Eskimo girl coming of age. 32 pages, hardbound, 10" x 8 3/8". (Ages 8+)

Price: $15.95 [members, $12.76], plus S/H (per book: U.S.–$2 book rate; $4 Priority Mail; foreign–$4 surface; $12 air mail)